THE
BODY IN
QUESTION

THE BODY IN QUESTION

Jonathan Miller

JONATHAN CAPE
THIRTY BEDFORD SQUARE
LONDON

For Rachel, Tom, William and Kate

Frontispiece
Detail from William Harvey's
'On the Generation of Animals', 1651.

First published 1978
© 1978 by Jonathan Miller
Jonathan Cape Ltd, 30 Bedford Square, London WC1

British Library Cataloguing in Publication Data

Miller, Jonathan
  The body in question.
  1. Physiology
  I. Title
  612   QP34.5

ISBN 0-224-01590-7

Printed in Great Britain by
Jolly & Barber Ltd, Rugby, Warwickshire

Contents

Acknowledgments

I would like to thank Karl Sabbagh of the BBC's Science Features Department, who invited me to compile the television series on which this book is based, and Patrick Uden who produced and directed the series. Throughout I have enjoyed the support and companionship of Jonathan Crane, Fisher Dilke and Jane Callendar, and I owe a special debt of gratitude to John Palfreman with whom I enjoyed many fruitful conversations. I have also received valuable advice and criticism from Eric Korn and Dr J. E. V. Bates of the National Hospital, Queen Square.

I wish to express my admiration and gratitude for the picture research provided by Robyn Wallis and for the programme material supplied by Jane Callendar. The difficulty of finding satisfactory illustrations was considerably eased by the enthusiasm and co-operation of Cliff Redman of the Royal College of Surgeons of England, the Science Museum, the Wellcome Museum of Medical Science and the Wellcome Institute for the History of Medicine. Ken Moreman, Gene Cox, the Department of Medical Illustration, St Bartholomew's Hospital, St Thomas' Hospital and Medical School, Brian Bracegirdle, the Department of Biology, University of York, and the Department of Chemical Pathology, Royal Postgraduate Medical School, Hammersmith Hospital. I also thank Susannah Clapp for her untiring editorial work, to which I should add a special note of gratitude to Anne Freedgood of Random House. Throughout the preparation of this book I have had the benefit of Ian Craig's tactful and expert advice on the design and layout of the illustrations.

I would also like to thank my wife and family for bearing with me throughout.

Finally, I wish to express my gratitude to the late Norwood Russell Hanson, who taught me philosophy while I was at Cambridge and introduced me to the interesting ambiguities of scientific observation. His tragic death while flying his own aircraft robbed the scientific world of one of its most interesting and provocative critics.

J.M.

Preface

This book has arisen because I was commissioned by the BBC to do a thirteen-part television series on the history of medicine. At the outset I was daunted by the prospect of making a chronological trudge from Hippocrates to Christiaan Barnard, and I knew that my own energy and patience would have been exhausted by the time I had reached the sixteenth century. In any case, writing history in this way presupposes that medicine steadily groped its way towards enlightenment and efficiency, its progress punctuated by flashes of genius and cries of 'Eureka!'. When I recalled my own medical training, however, I realised that the principles I had been taught and the assumptions which were supposed to guide my practice had their origins in the comparatively recent past, and that it was almost impossible to trace back a direct line of thought much beyond the seventeenth century. From that date on, the descriptions of the body and its processes are at least comparable with our own, and although the insights of the scientific renaissance did not have important practical consequences until the beginning of this century, it is possible to identify and sympathise with these founding interests.

Even so, medicine did not make an effective contribution to human welfare until the middle of the twentieth century. The great leap forward is often attributed to a rapid increase in heroic procedures and the discovery of new drugs, but what distinguishes the medicine of the past twenty-five years is not that its practitioners are equipped with an arsenal of antibiotics and antiseptics, but that they are furnished with a comprehensive and unprecedented understanding of what the healthy body is and how it survives and protects itself. We have today an impressive mastery of our illnesses precisely because we have a systematic insight into the processes which constitute health. This has been achieved by the accurate identification of the sort of thing our body is. And since finding out what something is is largely a matter of discovering what it is like, the most impressive contribution to the growth of intelligibility has been made by the application of suggestive metaphors.

In their efforts to manage and master the physical world, human beings have shown a remarkable capacity for inventing devices which lift, dig, hoist, wind, pump, press, filter and extract. With the use of furnaces, crucibles, ovens, hearths, retorts and stills, they have learned to transform the substances of the physical world into useful commodities. They have mechanised warfare and extended their powers of communication. The practical benefits of such ingenuity have been so impressive that it is easy to forget how much we have learned from the image of such mechanisms. While they have helped us to master the world, they have been just as helpful in giving us a way of thinking about it and about ourselves. It is impossible to imagine how anyone could have made sense of the heart before we knew what a pump was. Before the invention of automatic gun-turrets, there was no model to explain the finesse of voluntary muscular movement.

The immediate experience of the human body is something which we take for granted. We perceive and act with it and become fully aware of its presence only when it is injured, or when it goes wrong. Even then, the subjective experience of the body is usually incoherent and perplexing, and when we want it put right, we refer to people who have learnt to think about it with the help of technical metaphors: experts whose use of analogy has enabled them to visualise the body not merely as an intelligible system, but as an organised system of systems — which does not mean that man is an engine or that his humanity is a delusion.

It is unfortunate that the establishment of medical effectiveness has coincided with a large-scale rejection of scientific thought and with the identification of science as all that is destructive and unnatural in the human imagination. In the belief that modern man has deviated from the ancient wisdom of natural communities, many patients have turned their backs on orthodox treatment, favouring practices which they regard as the expression of some cosmic unity — homeopathy, herbal medicine, acupuncture, and so on. The irony is that far from rejecting or distorting nature, scientific medicine achieves its results by recognising what nature is and by reproducing and reconstituting her grand designs. Science is not a blasphemy; the wilful rejection of its insights is. In this book I have tried to show that one of the most effective ways of restoring and preserving man's humanity is by acknowledging the extent to which he is a material mechanism.

This is not intended to be a complete survey of human physiology: several organs and systems are conspicuous by their absence. I have tried instead to illustrate and explain some of the fundamental principles which constitute the basic assumptions of modern physiological thought: principles such as feed-back, self-regulation and the constancy of the internal environment. I have also tried to show how life maintains, defends, repairs and renews itself in a universe where there is a natural tendency to return to a state of uniform inertia and disorder.

But the book is not simply about the organs it describes, nor is it all that can be said about them. In fact, to some extent I have put questions about the human body in order to ask further questions about the nature of human thought, especially about the difficulty man has had in setting aside the notion that his body is worked by conscious mental processes. It is the story of the identification of the machine in the ghost.

J.M.

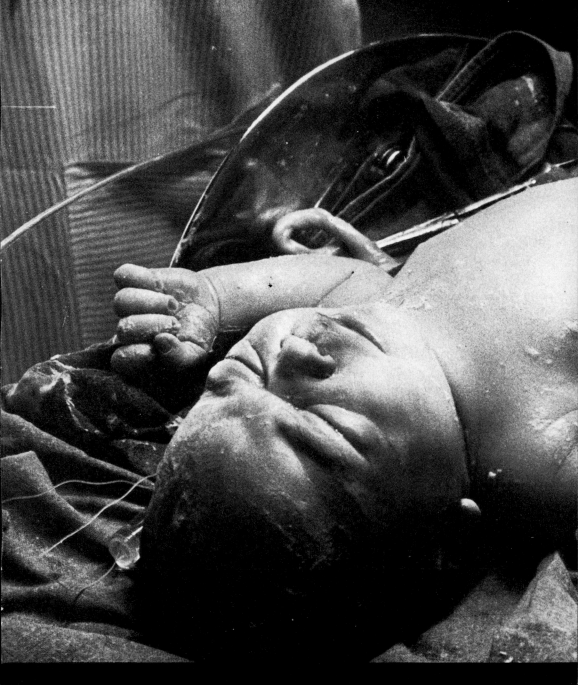

1 · Natural Shocks

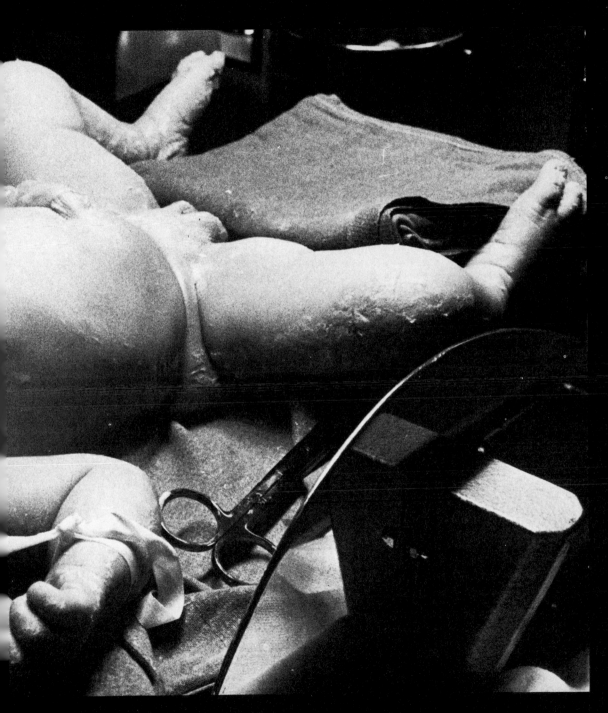

OF ALL THE OBJECTS IN THE WORLD, THE HUMAN BODY HAS a peculiar status: it is not only possessed by the person who has it, it also possesses and constitutes him. Our body is quite different from all the other things we claim as our own. We can lose money, books and even houses and still remain recognisably ourselves, but it is hard to give any intelligible sense to the idea of a disembodied person. Although we speak of our bodies as premises that we live in, it is a special form of tenancy: our body is where we can always be contacted, but our continued presence in it is more than a radical form of being a stick-in-the-mud.

Our body is not, in short, something we have, it is a large part of what we actually are: it is by and through our bodies that we recognise our existence in the world, and it is only by being able to move in and act upon the world that we can distinguish it from ourselves. Without a body, it would be difficult to claim sensations and experiences as our own. Who or what would be having them, and where would they be happening? Without a body, it would be hard to make sense of the notions of effort and failure and, since the concept of powers and their limits is built into the definition of personality, the absence of a physique through which these could readily be realised or frustrated would make it almost impossible to speak about the existence of a recognisable person.

The body is the medium of experience and the instrument of action. Through its actions we shape and organise our experiences and distinguish our perceptions of the outside world from the sensations that arise within the body itself. Material objects are called into existence by the fact that we can walk around them, get different views of them and eventually arrive at the conclusion that they exist independently of our own experience of them.

We can, however, also perceive our body as if it were one object among others. We can gaze at it, touch it, grope many of its contours, as if it were another of the many items in the world's furniture. Each of us, then, has two images of the bodily self: one which is immediately felt as the source of sensation and the spring of action, and one which we see and sometimes touch. In growing up, in emerging from the 'blooming, buzzing chaos' of infancy, these two images blend with each other so that the body which we see becomes the

On the previous pages The possession of a body may be the necessary condition of being a person but it is not a sufficient one. This new-born infant will have to learn how to become a person, taking actions which will eventually teach it to distinguish between self and non-self.

visible manifestation of the one which we immediately feel. Nevertheless, a moment's introspection will show how different these two images actually are.

When you close your eyes and try to think of your own shape, what you imagine (or, rather, what you feel) is quite unlike what you see when you open your eyes and look in the mirror. The image you feel is much vaguer than the one you see. And if you lie still, it is quite hard to imagine yourself as having any particular size or shape. Once you move, once you feel the weight of your limbs and the natural resistance of the objects around you, the felt image of yourself starts to become clearer, almost as if it were called into being by the sensations you create by your own actions — like a brass rubbing.

The image you create for yourself has rather strange proportions: certain parts feel much larger than they look. If you poke your tongue into a hole in one of your teeth, the hole feels enormous; you are often startled by how small it looks when you inspect it in the mirror. The 'felt' self is rather like the so-called anamorphic pictures with which artists entertained themselves in the Renaissance. The most famous example is the strange object hovering like a flying saucer in the foreground of Holbein's *The Ambassadors*: it is actually a splayed-out skull, which becomes immediately recognisable as such when viewed from the right angle. During the seventeenth century artists became very skilled at creating these transformations, which, if you place a cylindrical surface on the canvas, are at once restored to their normal proportions. So it is with the felt self and the visible self.

But although the felt image may not have the shape you see in the mirror, it is much more important. It is the image through which and in which you recognise your physical existence in the world. In spite of its strange proportions, it is all one piece, and since it has a consistent right and left and top and bottom, it allows you to locate new sensations as and when they occur. It also allows you to find your nose in the dark, scratch itches and point to a pain.

If the felt image is impaired for any reason — if it is halved or lost, as it often is after certain strokes which wipe out recognition of one entire side — these tasks become almost impossible. What is more, it becomes hard to make sense of one's own visual appearance. If one half of the felt image is

wiped out or injured, the patient ceases to recognise the affected part of his body. He finds it hard to locate sensations on that side and, although he feels the examiner's touch, he locates it as being on the undamaged side. He also loses his ability to make voluntary movements on the affected side, even if the limb is not actually paralysed. If you throw him a pair of gloves and ask him to put them on, he will glove one hand and leave the other bare. And yet he had to use the left hand in order to glove the right. The fact that he could see the

Holbein's anamorphic skull. ('The Ambassadors', National Gallery, London)

16

ungloved hand doesn't seem to help him, and there is no reason why it should: he can no longer reconcile what he sees with what he feels — that ungloved object lying on the left may look like a hand but, since there is no felt image corresponding to it, why should he claim the unowned object as his?

Naturally he is puzzled by the fact that this orphan limb is attached to him, but the loss of the felt image overwhelms that objection, and he may resort to elaborate fictions in order to explain the anomaly, fictions which are even more pronounced if the limb is also paralysed. He may claim, for example, that the nurses have stuck someone else's arm on while he wasn't looking; he may be outraged by the presence of a foreign limb in his bed and ask to have it removed; he may insist that it belongs to the doctor, or that prankish medical students have introduced it from the dissecting-room; one patient insisted that his twin brother was attached to his back.

When one half of the body image is eclipsed in this way, the patient frequently has difficulty in acknowledging or making sense of the corresponding half of the outside world. He finds it difficult, for instance, to draw symmetrical objects such as daisies or clock faces, and tends to crowd all the petals or

The partnership between 'felt' and 'visible' self.
('Cylinder Anamorphosis of Charles I' (artist unknown), Swedish National Portrait Collection, Gripsholm)

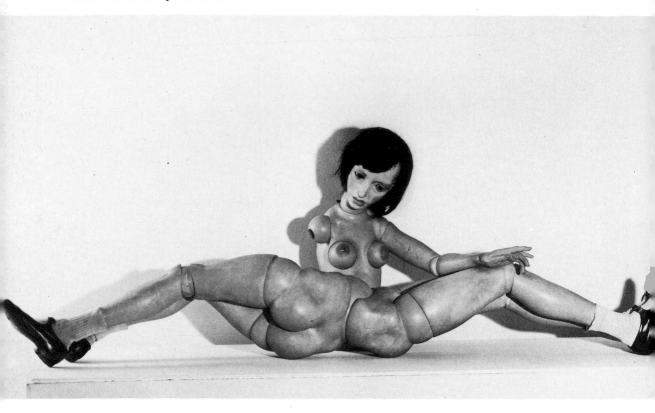

numerals on to one side. Such a patient may be able to tell the time between noon and six but be quite unable to read the hours between six and midnight. It is hard for him to find his way around the hospital, since he can appreciate turnings in only one direction and seems quite oblivious of the other. It is as if the world itself had suffered a partial eclipse.

An intact body image is an essential prerequisite for a full understanding of the shape of the world; which is not altogether surprising. The most inescapable experience we have is the sensation of our bodily self, and it is only in the course of growing up and acquiring skilled movements that we learn to tell the difference between the part of the world that is us and the part that is outside us. And just as the shrinking Roman Empire left Latin relics in the place-names of modern England — Manchester, Chester and Chichester — we leave linguistic remnants of our infant fantasy and label the world as if it were a huge body: hills have feet and brows; clocks have faces and hands; chairs have arms and legs; the sky frowns and the bosom of the ocean heaves.

In spite of its anamorphic proportions the 'felt' image of the body retains a regular coherence. Against this background of normality artists can play jokes and games. (Doll by Hans Bellmer)

Opposite   *In fevers and on the edge of sleep the body image can sometimes become exorbitantly distorted. (© André Kertèsz)*

18

Very occasionally, a patient appears to lose not just half but the whole image of his felt self, and it is then impossible for him to identify any sensations as his own. Mrs Gradgrind's death in *Hard Times* is a wonderful example of this:

'Have you a pain, mother?'
'There's a pain somewhere in the room, but
I cannot be certain that I have got it.'

Before you can recognise that a sensation is yours — before you can claim it and regard it as something that has happened to you rather than to the world at large — there has to be a felt self where it can be housed. Sensations happen in a rather strange part of the world, so strange that, strictly speaking, they don't happen in the world as such — at least, not in the way that explosions happen — but in an isolated annexe called the self, and if that annexe is missing or halved the sensations float around in a sort of elsewhere. If you have a ring on your finger and your hand is resting on the table, it makes perfectly good sense to say that the ring is resting on the table too. But if you have a pain in your hand and your hand is resting on the table, it sounds very odd to say that the pain is on the table as well. Pains don't happen in hands or heads or anywhere physical; they happen in the images of heads or hands, and if these images are missing the sensations are homeless. The reason we talk so glibly of having pains in our heads or in our hands is because under normal circumstances the subjective image of these parts coincides with their physical existence.

This situation can be reversed: the patient can lose a limb, and retain the image of it. Patients who lose legs and arms as a result of surgery or accident often report the feeling of a 'phantom' limb. They know that the physical limb has vanished, and when they *look* they can see that it is no longer there. Nevertheless, they *feel* an image of it, and they may even have phantom pains in it. The phantom limb may seem to move — it may curl its toes, grip things, or feel its phantom nails sticking into its phantom palm. As time goes on, the phantom dwindles, but it does so in very peculiar ways. The arm part may go, leaving a maddening piece of hand waggling invisibly from the edge of the real shoulders; the hand may enlarge itself to engulf the rest of the limb.

These phantom limbs are a painful ordeal, and surgeons are

often frustrated in their attempts to abolish them. It used to be thought that the sensation arose from the irritated ends of the wounded nerves, and surgeons used to cauterise these, generally to no avail. In fact, you can pursue the phantom to its source in the depths of the central nervous system, and still it persists. It is as if the brain has rehearsed the image of the limb so well that it insists on preserving the impression of something that is no longer there.

If the felt image of the physical self is in the nature of a fiction, an imaginary space which is usually occupied by the body of which it is supposed to be an imitation, where is this image housed? Where is the fiction created? If a surgeon opens the skull of a conscious patient and lightly stimulates the surface of the middle part of the parietal lobe, asking the patient to report what he feels, the patient will not mention or complain about sensations at the site of the stimulus. Instead, he will report strange tinglings in various parts of his limbs. As the needle is moved about, these sensations will alter their positions accordingly. By laboriously testing point after point, you find that the body is mapped on to the surface of the brain. It is the nervous activity of this map that creates the three-dimensional phantom we have of ourselves.

The brain map is not drawn to scale. Certain parts of the body are represented over a much wider area of brain than others, and not necessarily in proportion to their size. The face, especially the mouth, is allocated much more room than the leg; the hand, and especially the thumb, seem to have more than their fair share of space. It is like an electoral map as opposed to a geographical one. Because of their functional importance, the hand and the mouth have more sense organs per square inch than the leg or the trunk, and, since all of the parts of the body are clamouring for attention, they have many more Members representing them in their Parliament; that is to say, in the brain. This is what accounts for the strange anamorphic appearance of the felt image. The image that we see in the mirror reproduces the anatomical proportions of the body, whereas the image that we feel reproduces its Parliamentary proportions.

The electoral map is not a *picture* of the body, it is a neurological *projection* of it; that is, it is not painted on the surface of the brain, but called into existence through the

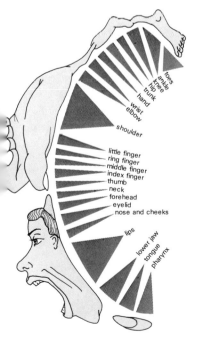

Penfield's famous 'homunculus', showing the proportional representation of bodily parts on the surface of the cerebral cortex.

nervous connections it has with the part it represents. We feel pain in the appropriate part of our felt image because there is a line of nervous connections between the sense organs in the skin and muscles and the Parliamentary representative in the brain which answers for each. If you trace the nerve fibres leading from the skin, you find that they join up, forming larger and larger cables as you go from the hand towards the shoulder. These cables enter the spinal cord in an orderly series of entrances. They then turn upwards and, as they make their way towards the brain, they form great bundles which grow still more as they are joined by new ones entering at each level. Within these bundles, the nerve fibres preserve an anatomical pattern: nerves from the leg are grouped near the centre, with nerves from the arm, neck and face joining them on the outside. At the top of what is called the brain stem, the sensory fibres are all collected and squeeze together, rather like the separate beams of light passing through a projector lens. After going through a section of the brain called the internal capsule, they spread out again and project themselves on to the surface of the brain.

Certain important parts of the body — the heart, the liver, the kidneys — are conspicuous by their absence from the brain map. Although the map is three-dimensional, it appears to be hollow, with nothing inside: it is as if a large part of the working population had no Parliamentary representation. This is why we have no felt image of the heart or the liver. The conscious relationship we have to our internal organs is rather like the one which brain-damaged patients have to their limbs: we may know that we have a heart, we have been told that we have a liver — but there is no felt image corresponding to them. Of course, patients with heart disease feel pain, and, as anyone who has had a kidney stone knows, you can get pains from the kidney; but we don't feel the pain *in* the heart or *in* the kidney, because there is no felt image in which to have such sensations. All such pains are referred: they are felt by proxy in a part for which there is already a felt image, and for each organ the proxy is always the same. The pain of coronary heart-disease, for example, is felt across the front of the chest, in the shoulders, arms and often in the neck and jaw. It is not felt where the heart is — slightly over to the left.

The reason that other internal organs consistently choose

Below and opposite
Serial segmentation in the earthworm. This elementary scheme is modified as evolution progresses but traces of the scheme can still be found in man, especially in the lay-out of the central nervous system.

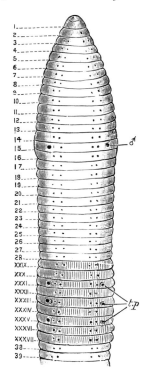

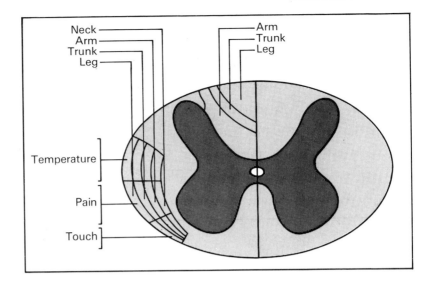

Cross-section of spinal cord, showing the orderly arrangement of sensory fibres ascending towards the brain.

the same Parliamentary proxy lies in the embryological origin of the organ in question and the fundamental architecture of the vertebrate body. Man and his vertebrate ancestors descended from a common stock and shared a basic plan. If you look at the earthworm, you can see that it is pleated at regular intervals from head to tail. This is not a surface ornament. When you open the worm, you find that the animal is arranged in a longitudinal series of segments, in each of which certain organs are repeated with monotonous regularity. In each segment, for instance, there is a pair of kidney tubes and a paired nerve supply branching off right and left. This structure is laid out at an early stage in foetal development, and the pattern is repeated in all creatures which have descended from this line of ancestors.

In fish, the chevrons of muscle correspond to the serial segments of the earthworm, and if you open the spinal cord you can see that the segmental pattern is repeated in the orderly sequence of nerves. In the higher vertebrates, this arrangement has been extensively remodelled, and it is often hard to detect signs of it without making a very careful dissection. Segments coalesce and the component parts are often reshuffled to adapt the body to the life of the individual creature. For example, the wing of a bird and the forelimb of a horse are both derived from the same embryonic segments.

The nervous system, however, often preserves the ancestral

23

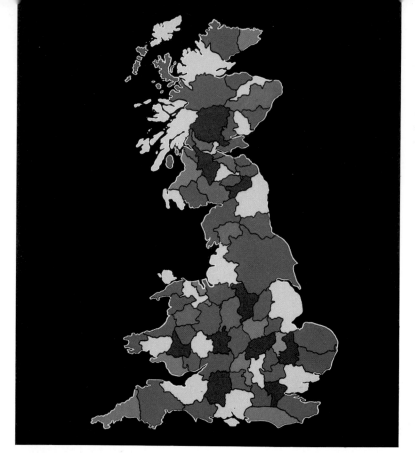

*Two alternative maps of the
British Isles. One reproduces
the metrical proportions of the
country, the other represents
the parliamentary proportions.*

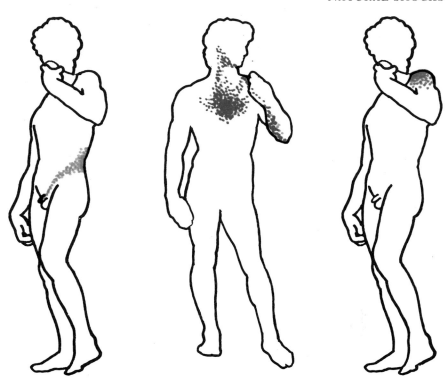

Kidney stone   *Angina*   *Ruptured spleen*

pattern. Nerves exit from the spinal cord in an orderly ladder
and, although these cables run together, divide and rejoin, it is
still possible to map their segmental distribution among the
skin and muscles. During the First World War clinical
neurologists compiled an atlas of segmental territories by
studying gunshot wounds in the spinal cord. By charting the
losses of sensation which followed known injuries to certain
nerve roots, they were able to draw up a territorial diagram
which they discovered preserved the old segmental pattern of
simpler vertebrates.

In man, the nerve segments which together form the neck
and the arms are also the ones where the heart appears. The
result is that the nerves bringing sensations from the heart are
in the same segment as the nerves which bring sensation from
the neck and arm. This relationship is preserved despite the
fact that in the course of foetal development the heart mi-
grates to a position which is quite remote from its original site.
It sinks down through the neck into the thorax and comes to rest
on the diaphragm, whose muscles are also derived from the
neck segments. But the heart maintains its ancient Parliamen-

25

tary representation, despite its position in the body: the neck, arm, and upper chest continue to feel the pain for it. The same form of representation applies to all those parts which one would loosely call the 'innards'. The spleen, like the heart, develops from the same segments which give rise to the neck and upper arm; when someone injures this organ in a football accident he frequently feels the pain at the tip of the left shoulder. An ulcer on the back of the tongue may refer its pain to its old segmental partner in the ear. As a kidney stone travels down the ureter it refers its sensations one after another to its old segmental sites: the pain characteristically pursues a long spiral course from the loin, round the side, and down to the top of the penis. Such pains are archaeological reminiscences of what we once were.

Just as our normal experience of the body is divided into two provinces, so when something goes wrong the symptoms occur in one or the other. There are immediate, self-evident discomforts or disabilities, which one can have only by noticing them; and there are the changes which have to be observed either by sight or touch (though one can exhibit them without necessarily recognising them oneself).

The second group, which one might call findings or discoveries, includes all the possible changes in complexion — pallor, jaundice, blueness, rashes, spots, eruptions; all changes in size, shape and weight — general wasting, local swelling, enlargements and shrinkages; and changes in facial appearance, such as staring eyes or drooping eyelids. These are in a sense the public features of illness: they are noticeable to everyone.

It is this conspicuousness that sets them apart from feelings or sensations which can be known only by the person who has them. Pain is a private experience, so is nausea, so are hunger and thirst. There are public *signs* of these states — groans, frowns, writhings, and so forth — but the actual pain and nausea and hunger and thirst are locked up in the unfortunate sufferer. The person with jaundice has only to exhibit it; someone with a pain has to announce it. Furthermore, having pain and knowing you have it are one and the same thing. If someone insisted he had a pain he couldn't feel, we would say that he had not learnt to speak English properly.

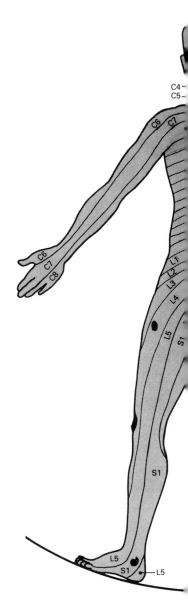

The human body betrays few visible signs of 'segmentation', but the distribution of spinal nerves bears witness to our ancestry. Diseases such as shingles follow the course of the sensory nerves and the rash paints a crude picture of one segmental territory or 'dermatome'.

26

On the following
pages *Disease may manifest
itself as a 'finding' rather than
as a feeling – as an exhibitable
change in the body's
appearance. Some of these may
be accompanied by feelings,
though – the swellings may be
painful and the red eyes may be
sore.*

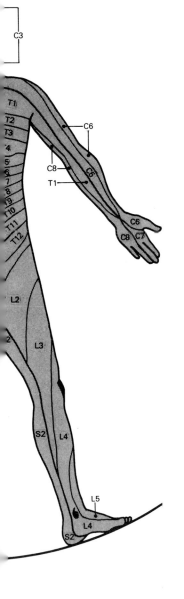

Sensations or feelings are also distinguished by the fact that there is no intelligible answer to the question 'What do you recognise them with?' You recognise swellings or rashes with your eyes, but you don't recognise pain with anything. It is obvious that a sense organ must be involved at some point in the proceedings. Why aren't we aware of this?

The answer is that the sense organs involved are very small and inconspicuous. The nerve endings which register these sensations are embroidered like millions of seed-pearls throughout the fabric of our body. With the help of a microscope you can find them in the skin, in the muscles and ligaments, in the walls of the blood vessels and in the membranes that line the abdominal cavity. If you link up their nerve fibres to an electronic recording device, you will see that they are constantly registering changes in their immediate environment, but they are much too small to be seen with the naked eye, and they are tucked away in inaccessible places.

This, however, is only part of the explanation. The fact that a sense organ is visible is much less important than the fact that we can control its performance, choosing and influencing the sensations we obtain. What makes us appreciate the visual function of our own eyes is not the fact that we can see that they are on either side of our nose, but that we can choose what we see *through* them: we can shade our eyes and reduce the glare; we can screw them up and blur what we see; we can close them and extinguish sight altogether; we can swivel our gaze this way and that, enlarging and exploring the visual field. Our ears are much less manoeuvrable than our eyes, but we can still muffle and sharpen our hearing at will; we can locate noises by rearranging the attitude of our head. If these actions are eliminated by, for example, listening to music on an expensive set of headphones, the experience of sound begins to resemble an internal sensation: the fact that one is unable to influence the balance and the volume by one's own actions means that the sound appears to come from somewhere in the middle of the head — the sound is not heard so much as had.

Of all our external senses, touch is the one in which action plays the most important part. Although the nerve endings are invisible, the fact that they are mounted on a mobile surface means that we can choose how and when they will be

27

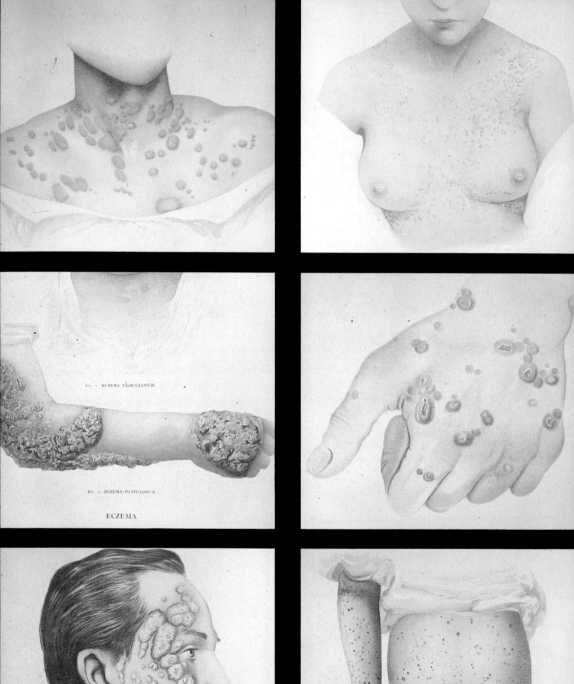

Fig. 1.—ECZEMA VESICULOSUM.

Fig. 2.—ECZEMA PUSTULOSUM.

ECZEMA

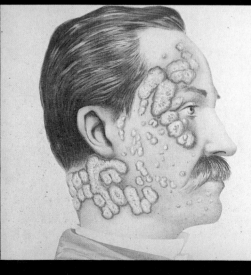

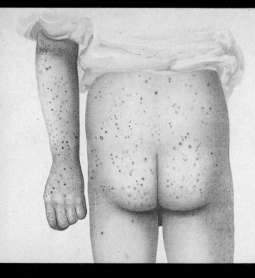

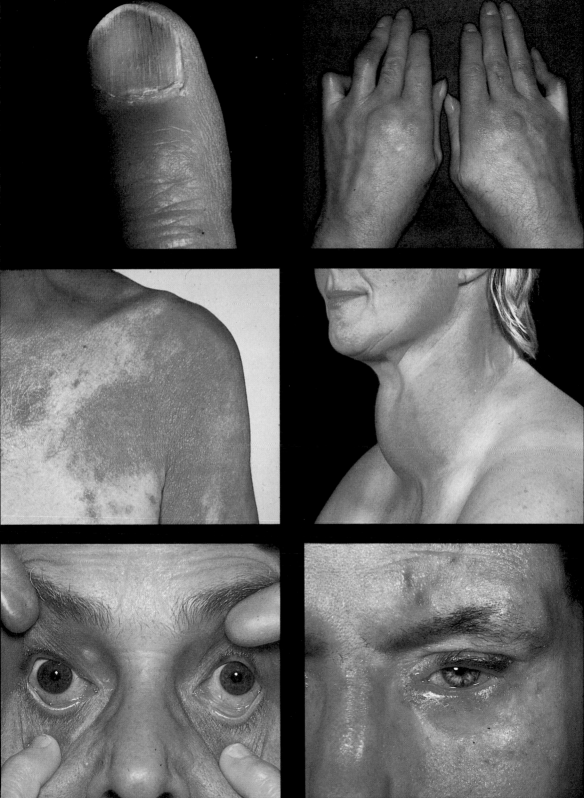

By correlating the sensory results of motor actions the infant learns to distinguish a world of permanent objects.

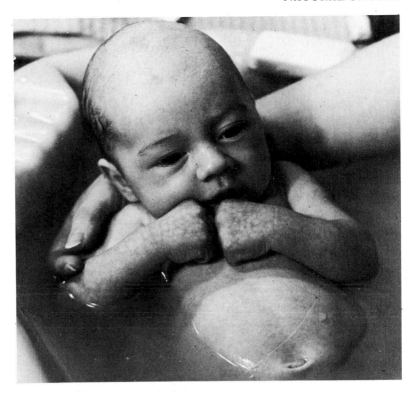

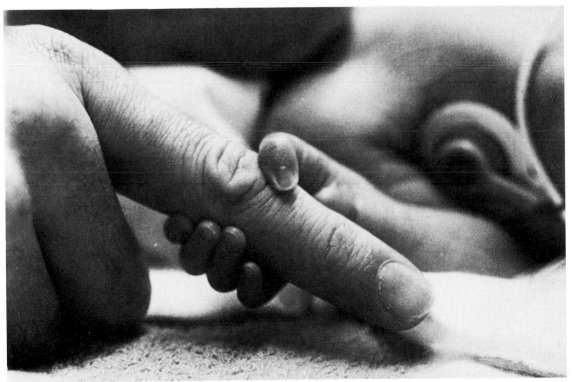

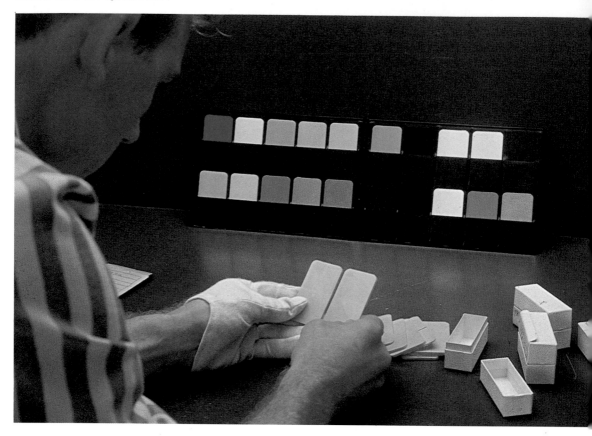

stimulated. We can squeeze them into action by gripping, stroking and poking; instead of having to wait for the world to touch us, we can reach out and obtain sensations when we choose. If we were completely immobilised — if all our touch sensations were imposed on us, and we were unable to take any action either to obtain or avoid them — the experience would be almost exactly like that of internal pain.

Although it would be difficult to give an intelligible answer to the question 'What do we feel pain with?', it would be even harder to make sense of someone who claimed to have a pain but was unable to say where it was. As we have seen, sensations are sometimes referred away from the place where the mischief is, but they are always located somewhere: they have a distinct and characteristic site, even if it is misleading. Naturally, some pains — those we call 'pricks', 'stabs' and 'shoots', for instance — are more sharply located than others. Those we call 'aches' tend to have a woolly, voluminous

The experience of vision is just as private as that of pain — but it can at least be discussed with reference to public standards. There are no 'samples' for comparing a given experience of pain.

feeling without any sharp outline. But even the vaguest pain can be pointed to, although you may have to use your whole hand rather than the tip of your finger. In fact, a lot can be learnt about a pain from the way in which the patient points to it. Apart from saying something about where it is, the movement of the hand is often a tell-tale sign of its quality: if someone has angina, he often presses the front of his chest with a clenched fist; the whole fist shows that the pain is widespread; the fact that the hand is clenched tells us the pain has a gripping quality. The pain from peptic ulcer is often closely localised, and the patient usually tells you so by delicately pointing to it with the tip of his index finger. When a stone lodges in the outlet to the kidney, the pain, as we have seen, tends to radiate in a long oblique line from the loin to the groin. Even if the patient doesn't describe this movement in words, he will sometimes do so by holding his side with the thumb at the back and the fingers pointing down at the front. If a pain is more or less superficial — what would normally be called soreness — the patient may lightly brush the surface of the skin with the outstretched tips of all five fingers. This sometimes happens in the early stages of shingles. The skilled physician can learn a lot from the pantomime of complaint.

Although it is hard to imagine an unlocated pain, there are strong sensations or feelings which are not recognisably associated with a part of the body. It is very hard to say where you feel nausea, for example. If you ask people, they often simply say that it is what they feel when they know they are going to be sick. But what about people who have never been sick? What do they feel?

The problem of location becomes even more puzzling with experiences like hunger or thirst. Thirsty people usually complain about a dry mouth, but if that were all there was to it, the feeling of thirst would be wiped out at the first mouthful of water, and it isn't, as a very elegant experiment has proved. If the gullet of a dog is taken out through its neck and the dog is made thirsty, it will continue to gulp water incessantly for twenty-four hours. It never relieves its thirst. The same difficulty applies to hunger: although there are abdominal pangs associated with *extreme* hunger, the sensation of craving for food has no definite location. We now know that thirst and hunger are due to changes in the chemistry of the blood and

33

that these are registered by sense organs in the brain. However, one doesn't feel hunger either in the blood or the brain —
one simply feels it, and that is that.

Up to this point I have dealt with two of the major categories
of pathological experience: findings and feelings. Now we come
to the third and last category, that of experiences which can
be conveniently grouped under the single heading 'failings';
that is, errors of performance. Some of these are experienced as if they were feelings, but do not fit quite comfortably into that group: numbness, for instance, blurred vision,
and hardness of hearing. Most people would call a numbness a
sensation, but unlike pain it is not really a sensation in its own
right: it is the feeling you get when the normal feeling of touch
is not working properly. If you think of the body as a map,
pain is like a stain on the surface of the map, and numbness a
drop of bleach. In the case of pain you are adding a feature to
the map; in the case of numbness you are removing the
possibility of it happening. This also applies to blurring of
vision, which you can have only by trying to see normally. It is
a comparative sensation — just as deafness is. For that reason,
numbness, blurring of vision and hardness of hearing should
be regarded as failures of performance, more closely related to
symptoms like paralysis and tremor than they are to sensations like pain or itch. Although paralysis is an observable
sign, the patient experiences it as an unexpected failure of part
of what he regards as his normal repertoire of powers.

*Numbness is the feeling that
accompanies the absence of
feeling. It is a symptom of
subtraction.*

   All errors of performance — failures of sense organs on the
one hand, and straightforward failures of mechanical power
on the other — are comparatively simple. But human performance can fail in much more elaborate ways, on both the
sensory and the executive side. Patients with serious forms of
brain injury often show dramatic errors in their sensory
performance, in spite of the fact that their eyes, ears and hands
are intact. If a patient has received an injury to that part of the
brain which classifies, judges and criticises the information
provided by the eyes, he will fail to recognise an object which
is presented to him in a visual form, although he promptly
identifies it when it is placed in his hand. With another type of
injury, the patient is quite unable to make sense of the familiar
object placed in his hand, but accurately names it as soon as he

opens his eyes. Such failures of recognition form a large class of illnesses which go under the general name of agnosia — failures to make sense of what the senses provide. A comparable failure, known as apraxia, occurs when the link between the understanding of the request and the action that fulfils it has been broken. Patients whose muscles are in perfect working order — who automatically groom and dress themselves — fail to obey simple orders although they appreciate what they have been asked to do and have the muscular wherewithal to do it.

Language can also fail as a result of disease, and the character of the failure depends on whether it breaks down at the level where it is uttered, or at the point where the utterance is conceived. A local paralysis of the vocal chords may render someone completely silent, but his mastery of language remains unimpaired: he can still convey his thought by means of writing and make graceful and eloquent use of sign language. The same is true of disturbances of articulation: speech may be slurred, staccato or, as in certain cases of multiple sclerosis, may acquire a strange syllabic rhythm — so-called 'scanning' speech — but the patient's linguistic competence remains intact. However, if the damage occurs at a higher level, as it does in strokes, the patient may suffer from one of the various forms of aphasia, the inability to express himself linguistically. The neuro-muscular apparatus of the voice and tongue may be intact, but he cannot put his thoughts into words. By examining the brains of such patients at post-mortems, neurologists discovered that this happens when right-handed people suffer a haemorrhage on the left side of the brain — and vice versa — or, in other words, that the linguistic centres are localised in one particular part of the brain.

Memory, too, may fail as the result of brain damage. Acute head injuries, for instance, can rob the patient of any memory of the events immediately leading up to the accident. A crash victim may recall everything up to the point where he opened his garage door in the morning, but be quite unable to remember what happened next. Or the defect may be much more diffuse. Slowly developing brain damage, due to arteriosclerosis or chronic alcoholism, can cause a widespread impairment, and the patient may find it impossible to retain the simplest facts: engagements, dates, names and ideas; he may

Pain is a positive sensation added or superimposed upon the matrix of normal feelings.

not know where he is or even who he is. In some cases he may try to cover this up by inventing florid and plausible details — a combination of amnesia and confabulation which is known as the Korsakov Syndrome. Although some neurologists insist that the confabulation is just another aspect of the patient's mental confusion, there are others who claim that it illustrates a natural tendency of the brain to complete or fill in unaccountable gaps in experience.

Failures of performance are not confined to neurological diseases. A person's performance can falter in a large number of alternative ways. Under normal circumstances, for instance, one expects to be able to empty one's bladder as soon as it is felt to be full. The evacuation starts promptly, continues fluently and finishes conclusively — and one doesn't expect it to leak in between. However, when the outlet from the bladder is partially obstructed, as it sometimes is by an enlarged prostate, a patient's ordinary performance begins to fail: he may have difficulty in starting, the stream is feeble and hesitating, and he is unable to finish off, so that he continues to dribble. A normal function has become inefficient, and it is this inefficiency which makes him complain. Conversely, the failure may be one of containment rather than evacuation. A normal bladder can retain its contents even when the pressure is high, but if the muscular outlet is weakened, as it sometimes is after a mismanaged or strenuous childbirth, the patient may suffer from stress incontinence. Coughing, sneezing or laughing will overcome the normal efficiency of the intact sphincter, and the patient will wet herself.

The trio of findings, feelings and failings covers most of the ways in which disease can make itself known, but there are certain episodes which don't fit comfortably into any of these three categories that might be mentioned briefly. One would hardly, for instance, call vomiting a 'finding'; and, although it is preceded by a feeling and accompanied by some unpleasant sensations while it is going on, it is quite clearly an incident in its own right. On the other hand, one wouldn't want to call it a failure of performance either, since it is not something which one could do better. It *is* a performance, a physiological competence which accompanies illness and represents the body's attempt to overcome it. As we shall see in a later chapter, we should be in trouble if we couldn't vomit. The same principle applies to coughing and diarrhoea, which, like vomiting, are attempts on the part of the body to rid itself of irritants, infections or poisons.

Sudden haemorrhage is different. It may present itself as a finding, it can result in a feeling, but one can't really regard it as a failure of performance. Perhaps one should simply regard it as an eruption — something which bursts upon the scene, creating anxiety and terror.

One of the most peculiar experiences is convulsion. Although such episodes usually originate in the nervous system, they are not failures of performance: they may interrupt performances and in a sense even be performances. Nervous tremors or tics vitiate the movements which they accompany, whereas convulsions interrupt the flow of life by compelling the patient to do things he would prefer not to: perhaps we should call them 'compulsions', or 'brief tyrannies'.

These, then, are the experiences which make up the concept of illness: the 'natural shocks that flesh is heir to'. The attention that is given to them, however, and the actions to which they give rise depend on the person to whom they are happening. A large, unsightly lump may pass unnoticed by one person, while a small swelling can cause great alarm in someone else; a pain which is played down by one person can cause another to cry out; a vain man with a well-lit mirror will often recognise the first tinge of jaundice, whilst a more complacent person might have to turn bright yellow before he became aware of a colour change. Some people are simply more stoical. People who have been trained to look out for abnormal signs — such as medical students — tend to be hypochondriacal. Fear also plays an important part both in alerting and in blinding people to the possibility of illness: someone who is over-anxious about his health tends either to recognise signs and symptoms which his neighbour would overlook altogether, or to avoid examining himself for fear of finding out the worst.

The patient's command of language is also influential in the experience of illness. Patients with a small vocabulary may be at a loss to describe their feelings, even to themselves — and when someone hasn't got a name for something, it is much easier for him to neglect or forget it. But even when the patient is comparatively articulate, the doctor can never be sure just how he is using words. It is particularly hard when it comes to feelings — aches, pains and itches — where the witness has no way of sharing the experience. One of the ways in which we try to overcome this problem is by mentioning the type of accident or injury which might produce a pain of that sort. We talk about stabbing pains or scorching pains or heartburn, or we say that a pain is like being squeezed in a vice.

But there are drawbacks to using language in this way. How many people have actually been stabbed? It is all very well to try to imagine what it might be like, but if neither of us has ever been stabbed, I may imagine something completely different from you. In fact, it is surprising that there is as much agreement as there is: when people talk about stabbing pains they are usually referring to sensations which come and go with great speed and violence. Presumably the patient has abstracted from his *idea* of stabbing an image of violent penetration which he uses in a metaphorical way to refer to his own pain. Oddly enough, people who have actually been stabbed rarely talk about the pain in this way. And vice versa: the events which cause a so-called stabbing pain are not actually stabs. Similarly, when someone talks about a bursting headache you don't necessarily expect to find something bursting inside his skull. In fact, when a blood vessel does burst inside the head, patients usually describe the pain as a blow rather than an explosion: 'I felt as if I'd been struck on the back of the head by a bat,' they often say.

So the quality of the pain is not always a reliable picture of what is happening inside, just as the place where the patient feels the pain is not necessarily the seat of the mischief. In fact, compared to our knowledge of the external world, we have a very limited acquaintance with our own physique. Close as it is, we know much less about it than about anything else. In the normal course of events all we can feel is the vague mass of our head, trunk and limbs. As the day goes on, we forget about the clothes on our back, and when we do become conscious of our own surface it is usually because it has been assaulted by unexpected sensations. And the feelings we get are almost invariably ascribed to the outside events which are responsible: for instance, if you pick up a pencil, what you feel is an object of a certain weight, size and shape — you are not aware of the dimples in the tip of your finger. Exertion or effort can strengthen the feelings of joint and muscle, but no one is aware of the individual muscles involved in clenching one's fist.

This state of affairs is not just a regrettable accident, however: we cannot expect any improvement. Our nervous system is designed to emphasise what's going on in the outside world; our intelligence faces outwards, and our survival depends on the way in which we appreciate the threats and opportunities of the world beyond. It is as if our bodies formed a political state in which the Prime Minister or President takes over the portfolio of the Foreign Secretary, leaving the conduct of Domestic Affairs to a Home Secretary who works independently, supervising all the necessary changes and adjustments without having to refer to the highest office.

The inside of the body — blood vessels, heart, intestine, lungs and bladder — is literally studded with instruments capable of registering changes in pressure, temperature and chemical composition. But none of these meters has any dials: they are not meant to be read by human consciousness, but are linked up with the reflex systems which obey automatically. Although you can sometimes, when the variations are extreme, become aware of the adjustments going on inside, most of the time you are quite unconscious of the hectic activity. When you start to exercise, your heart automatically speeds up in order to supply the working muscles with the blood they need. But you don't make the decision to speed up your heart. You simply choose to run, and the reflex arrangements automatically take care of the rest. And when you stop running, you don't have to remember to switch off your heart: as the sensory messages stop arriving from the muscles, the heart begins to slow down automatically; one of the first signs that something is wrong with the circulation is that it takes the heart longer than normal to slow down after exercise.

This principle applies most of all to systems which have a closed cycle: the circulation of the blood, for instance, which follows an unbroken sequence without coming into immediate contact with the outside world; the blood vessels are constantly adjusting themselves, redistributing blood from the skin, the muscles and so forth. When you need to lose heat, the radiators of the skin flush with warm blood, whereas when you need to conserve heat, the blood vessels contract and the skin whitens and cools. You can sometimes feel the results of all this, but you are not actually conscious of the act of contraction or expansion. You can't blush at will.

Certain cycles, however, open directly into the outside world, and in these cases we are made aware of the sensations in order to introduce voluntary control of what happens. Take the urinary system. The kidneys manufacture urine from the blood, following biochemical instructions which never enter our consciousness. If the body is short of water for some reason, the urine is automatically reduced in volume and becomes more concentrated. If it is short of salt, the kidneys reabsorb as much as they can. The blood is continually being monitored or tasted by sense organs linked up to systems which do everything they can to maintain the *status quo*. All this is quite automatic: the urine sweeps into the bladder through the ureter, and, although there are muscles to help it on its way, we are quite unaware of their action. In animals which live in the sea, the urine can flow straight out into the environment without the animal having to pay any attention. Such carelessness isn't possible on dry land: an unsupervised flow of urine would soon ulcerate the skin, and before long, infections would backtrack into the ureter. Land animals have therefore developed a muscular reservoir for holding the urine until it can be thrown clear in one go. In lower animals, this act is more or less automatic, but in animals like cats and dogs the sensations of a full bladder arouse enough of their attention to cause them quite complicated behaviour, and in species which live in communities, like ours, the part played by consciousness is even more important. What happens is this: the urine accumulates unnoticed until the pressure begins to stretch small sense organs embedded in the wall of the bladder; in an untrained infant, these impulses bring about an automatic reflex; in an adult, the sensations rise into consciousness, and etiquette takes care of the rest. In patients suffering from a broken back, the messages between bladder and brain are interrupted, and their urine accumulates until it reaches a critical pressure, at which point the bladder opens automatically, regardless of where the patient happens to be.

The same principle of dual control applies to the intestine, except that this system is open to negotiation at two ends.

Although it is automatic throughout its huge length, it still has to be filled and emptied at the right moment, and at both points this involves a conscious transaction with the world. The feelings of hunger and taste allow us to negotiate for the right meal, whereas the sensation of a full rectum reminds us when to visit the lavatory. Between these two archways of conscious sensation lies the vast unconscious Amazon of the intestine. All its movements are quite automatic. (Which is just as well — imagine what it would be like if you had to supervise the passage of food through its whole length.) Fortunately, the muscles of the intestine move of their own accord, and the peristaltic waves follow one another in an orderly fashion.

But we have to pay a price for this labour-saving efficiency. When anything goes wrong, neither the quality nor the location of the sensation tells us what is happening. A kidney stone, for instance, produces an agonising pain, but unless you have had it before and know that this is the sort of pain you get with a stone, you are almost bound to be mystified by it. In one way or another, this applies to all our innards — heart, lungs, intestines. In emergencies, the sensations and feelings may be extremely vivid and can in fact monopolise our attention altogether, but the feelings we get are almost entirely uninformative. Designed to work without conscious in-

You cannot experience your own interior by closing your eyes and concentrating on it. In order to discover your own contents you have to investigate the inside of someone else.

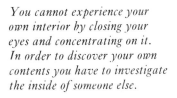

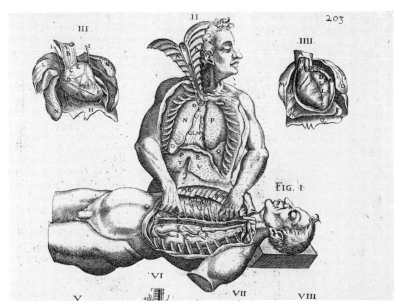

struction, these interior systems are at a loss for words when they try to speak up for themselves. And, to make it worse, the vocabulary of internal sensation is so small that several illnesses have to share the same feelings. Breathlessness may be the result of anaemia, pneumonia or heart failure. A bad internal haemorrhage with concealed loss of blood produces symptoms almost indistinguishable from a sudden fall in blood sugar. Nausea may arise from food poisoning, kidney failure, appendicitis, brain tumour or migraine.

Although our experience of our body is so vague and muddy, our mind does everything it can to intensify the images with which it is supplied — like the computers which sharpen the pictures sent from distant planets. In the absence of any immediate knowledge of our own insides, most of us have improvised an imaginary picture in the hope of explaining the occasional feelings which escape into consciousness. Our mind, it seems, prefers a picture of some sort to having to live through the chaos of sensations that would otherwise seem absurd.

To some extent, this applies to all our experience. One way or another, our senses introduce us to a world, and not just to a kaleidoscope of sensations. The visual world, for example, is not just a patchwork of tints — although that is all there is on our retina. A blotch of green on a rural horizon is usually seen as a tree. It may turn out to be something else — a painted barn or a group of camouflaged soldiers — and until the issue is settled one way or another we alternate between these interpretations. But we never stick half-way and see what is actually on our retina. In the very act of entering consciousness, sensations are somehow made up into scenery and we experience them as objects in a world. But this doesn't happen automatically. The mind has to make a guess about the identity of what it sees, hears or feels, and the odds are determined by all sorts of hints and hot tips. There is a well-known picture puzzle which shows this process very neatly. On page 43 is a picture of a group of courtiers. If I invite you to see a bust of Voltaire, the well-known features will suddenly jump into sharp relief. I have shifted the odds in favour of a famous face. By a strong effort of the will — whatever that means — you may be able to wipe out the face and restore the courtiers. But see how hard it is to hold your vision mid-way

Camouflage is a biological mechanism for shifting the perceptual odds in favour of an alternative visual interpretation.

The acknowledged existence of Houdon's famous bust of Voltaire makes it possible to 'see' Dali's picture as something other than a panorama of Spanish courtiers. What one perceives depends on what one knows.
(Houdon, Victoria & Albert Museum; Dali, 'Slavemarket with the disappearing bust of Voltaire', Salvador Dali Museum, Cleveland).

and see the picture just as it strikes your retina.

This principle also applies to our conception of our own internal world, although here the patterns are much less coherent and picturesque. We talk about feeling liverish or having heartburn or chills on the kidney, but all these visceral images have to be invented, and our only source of information for the inventions is what we have been told. We reconstruct our insides from pictures in advertisements for patent medicines, from half-remembered school science, from pieces of offal on butchers' slabs and all sorts of medical folklore.

These promptings may account for the otherwise inexplicable fact that the French seem to have far more trouble with their liver than the English do. It is hard to believe that this organ is so much more threatened in France than it is in England, in spite of what we have been told about their drinking habits. It seems more reasonable to assume that the

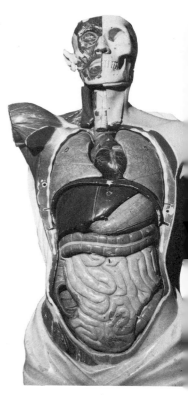

French interpret their symptoms in the light of a national fantasy about the liver and unconsciously reshape their sensations in terms of this phantom organ. The English, on the other hand, are obsessed with their bowels. When an Englishman complains about constipation, you never know whether he is talking about his regularity, his lassitude, his headaches, or his depression. Once an organ gains a hold on the collective imagination, its influence is almost invariably exaggerated, and a wide range of symptoms are explained in terms of it.

The experience of friends and neighbours also plays a large part in the editing of sensations. If someone we know has heart trouble, it is only too easy to reshape our own sensations until they come to resemble his. Most of us have scattered pains in and around the chest at some time or another, and, unless we are abnormally sensitive, we disregard them. But if one of our friends or relatives is known to have something

The merchandise displayed by the butcher may help to furnish an image of the human interior. In the same way half-remembered images from schoolrooms may help to shape the experience of internal disease.

44

wrong with his heart, these scattered sensations grow together and easily assume the shape and permanence of the rumoured pain.

Almost invariably it is the sufferer who shapes the experience of illness into an intelligible situation. At some level or other suffering gives way to a personal diagnosis. In the end the victim becomes a patient because he guesses that something is wrong or odd about him and that he is the unwilling victim of this process.

From the instant when someone first recognises his symptoms to the moment when he eventually complains about them, there is always an interval, longer or shorter as the case may be, when he argues with himself about whether it is worth making his complaint known to an expert. Naturally, it varies from symptom to symptom: someone who recognises that he is steadily losing weight measures this by a different standard from the one he would apply to a headache; the thoughts aroused by a painless swelling on the cheek are not the same as the intense irritation caused by an itching rash on

Commercial images also help to fill in the gaps. Advertising plays upon and then helps to exaggerate national obsessions.

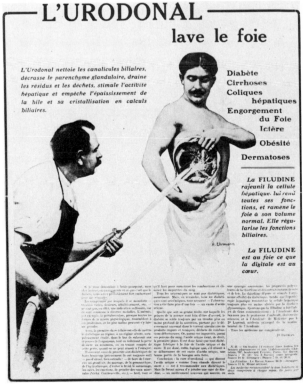

the hand; when someone eventually decides to complain about increasing breathlessness he has rated himself on a different scale from the one he would use if he had blood in his water.

There are four independent scales which can be applied to a symptom to decide whether or not it demands attention. First, there is the symptom's intrinsic nastiness. Pain is intrinsically nasty, and most people would agree that nausea is, as well. The straightforward feeling of having either of these symptoms is quite enough to make you wish you hadn't.

In contrast to these, there are symptoms whose nastiness has to be inferred. The painless appearance of blood in the urine is not obviously unpleasant as a thing in itself. It is unsettling not because it causes discomfort, but because it carries alarming implications. The same principle applies to a painless, invisible lump in the breast.

Symptoms are also complained about because they reduce efficiency or restrict freedom of action. A patient who becomes steadily more breathless when he exerts himself eventually complains because he is frustrated by the way in which this cuts down his movements. It is not the feeling of breathlessness as such, but the results of it which lead him to complain. This also applies to tremor or muscular weakness of one kind or another: unless they come on rapidly enough to cause immediate suffering or alarm, they excite complaint because of the frustration they cause.

Finally, there is the question of embarrassment or shame. Patients may complain about a symptom simply because they regard it as unseemly. A painless blemish on the mouth can bring a patient to the clinic much sooner than a large, painful lesion on the shin. This applies to any publicly noticeable anomaly — a squint, drooping eyelid, hairlip, rash, loud intestinal noises, hoarseness — which seems to threaten the patient's self-esteem and which might be regarded as a stigma.

Obviously, a symptom can appear on more than one scale — perhaps on all four. The pain associated with cystitis — infection of the bladder — is extremely unpleasant in its own right, so it scores quite high on the intrinsic nastiness scale. On the other hand, it poses no threat to life, so that although it may excruciate the patient it doesn't actually alarm her. However, patients who suffer from severe cystitis know only

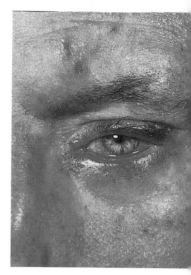

The nasty sensation of soreness may have brought this patient to the doctor – but it could just as easily have been the shameful unsightliness.

Shame can also be felt by proxy. A mother may feel discredited by the rash on her child's body and may complain to a doctor, not because the child is distressed but because she is embarrassed.

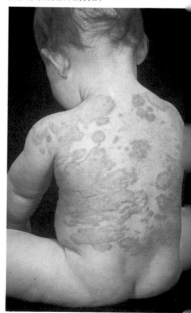

Acute Breathlessness

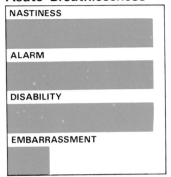

Senile Tremor

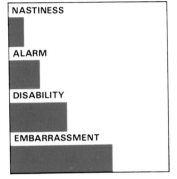

Cystitis

Hot Flushes

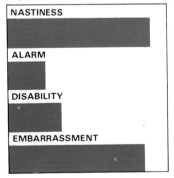

too well how much it limits their freedom of action: long journeys are almost inconceivable, and while the illness lasts the unfortunate patient tends to hover within easy reach of a lavatory. So it gets a high score on the incapacity scale, and since it alters the patient's behaviour in what she regards as a noticeable way, it also gets a substantial score in terms of stigma and shame.

Angina is a pain that is unquestionably nasty, and since most people have learnt to associate it with the heart they are often alarmed by it as well. But greater perhaps than the discomfort and alarm is the sheer frustration of being brought to a standstill after walking less than 100 yards. It may be exasperation that eventually drives the patient to a doctor.

Breathlessness is an even more interesting example. It is more or less unpleasant in itself, and if the patient associates the symptom with his heart or his lungs he may also be alarmed by it. But, as with angina, the sense of incapacity may weigh most heavily, and if the patient prides himself on being able-bodied he may experience an intolerable sense of shame when his friends or colleagues leave him panting on a short stroll. However, if it comes on very rapidly or very severely, the scale changes. Someone who is awakened at night by an attack of paroxysmal breathlessness finds the feeling almost insufferable in itself, so that it now hits the very top of the intrinsic nastiness scale. Since most people would interpret it as a serious threat to life, it notches up a heavy score in terms of sheer alarm as well. And when breathlessness is extreme, the patient is often bed-ridden, so it also appears very high up on the incapacity scale. However, anyone preoccupied with drawing his next breath has very little time to feel ashamed of it, so that it doesn't even appear on the stigma scale, although a sense of shame and embarrassment can survive even the most fearful emergency — I can remember patients struggling for breath, waving a weak apology for what they obviously regarded as humiliating panic.

Elderly patients often expect to have shaky hands, and if they develop a tremor, as long as they have no reason to suspect a sinister cause, they are simply frustrated, since it makes it hard for them to dress and feed themselves. But even elderly patients may be embarrassed by their tremor: they are ashamed of not being able to help themselves and also imagine

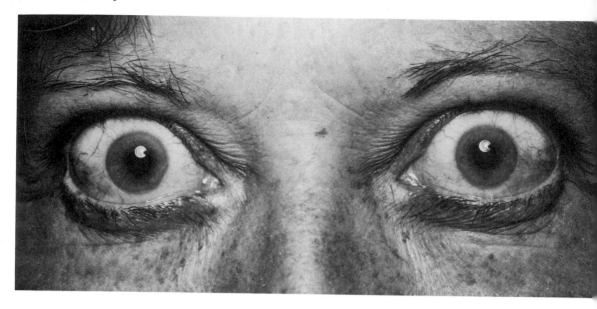

The patient with thyrotoxicosis usually complains about nervousness, tremor and intolerance of hot weather. But the embarrassment caused by the characteristic exophthalmos (or staring eyeballs) may outweigh all the other considerations.

In cultures where the display of the body involves a loss of 'face', doctors may have to resort to indirect methods in order to let the patient complain accurately. This Chinese figurine allows the patient to point to the affected part without actually showing it.

that people will think they are stupid or odd or even drunk. Once again the way in which this symptom is rated depends on other factors: its severity or the age of the victim at the onset. I can remember a patient who developed a spectacular tremor at the age of forty as the result of breathing in mercury fumes: this is the illness they used to call Hatter's Shakes, because of the mercury which was used to dress the felt. (I think it is probably the reason why the Mad Hatter in *Alice in Wonderland* is shown with a large chunk bitten out of his teacup.) My patient said that there was nothing intrinsically nasty about the shakes, and, since he had no reason to associate them with a dangerous industrial poison, he wasn't even alarmed by them. But he was seriously incapacitated by his tremor and what worried him most was the fact that it stopped him going to work — luckily for him, as it turned out. He was also embarrassed by his clumsiness.

To sum up, then. At one time or another we have all been irked by aches and pains. We have probably noticed alterations in weight, complexion and bodily function, changes in power, capability and will, unaccountable shifts of mood. But on the whole we treat these like changes in the weather: as part and parcel of living in an imperfect world. The changes they cause in our behaviour are barely noticeable — not inconvenient enough to interfere with our routine. We may retreat a little, fall silent, sigh, rub our heads, retire early, drink glasses of water, eat less, walk more, miss a meal here and there, avoid fried foods, and so on and so on. But sometimes the discomfort, alarm, embarrassment or inconvenience begin to obstruct the flow of ordinary life; in place of modest well-being, life becomes so intolerably awkward, strenuous or frightening that we fall ill.

Falling ill is not something that happens to us, it is a choice we make as a result of things happening to us. It is an action we take when we feel unacceptably odd. Obviously, there are times when this choice is taken out of the victim's hands: he may be so overwhelmed by events that he plays no active part in what happens next and is brought to the doctor by friends or relatives, stricken and helpless. But this is rare. Most people who fall ill have chosen to cast themselves in the role of patient. Viewing their unfortunate situation, they see themselves as sick people and begin to act differently.

Usually this is a prelude to seeking expert advice, but falling ill can sometimes be performed as a solo act. In New Guinea, for example, the decision to fall ill is almost invariably followed by a consistent and easily recognised form of behaviour. The sufferer withdraws from the community and retires into his hut: he strips himself naked, smears himself with ash and dust, and lies down in the darkness. He also changes his tone of voice: when his friends and relatives make solicitous enquiries, he answers them in a quavering falsetto.

Such people are not merely suffering illness: they are performing it, thereby announcing both to themselves and to the community that they are sick people in need of care and attention. In New Guinea this is such a well-recognised form of behaviour that one is tempted to regard it as a formal ritual. Something similar, however, can often be found in more sophisticated communities. When someone falls ill but is not yet ready to summon expert help, he usually takes care to advertise his condition through the medium of a performance. In fact, such a performance is often demanded of him by those with whom he lives. Someone who takes to his bed when he has a sick headache, for instance, is not entirely prompted by the need for relief. It is a way of boosting his credibility as a sick person, and it may be the only way of getting the attention and concern which he thinks he deserves. In fact, the patient may have to abstain from activities he is quite capable of performing, if only to convince those around him that there is a good reason for his staying away from work.

A patient, then, is a special sort of person, rather like a recruit or a convert or a bride. By taking on the role of patient you change your social identity, turning yourself from someone who helps himself into someone who accepts the orders, routine and advice of qualified experts. You submit to the rules and recommendations of a profession, just as a novice submits to the rules and recommendations of his or her chosen order. Ordinary life is full of such voluntary transitions — changes of social role or status which are accompanied by corresponding changes in obligation and expectation. Whenever these take place, they are accompanied by rituals which mark the event and make it clearly recognisable to all who are involved. The anthropologists have called these 'rites of passage', symbolic actions which represent and

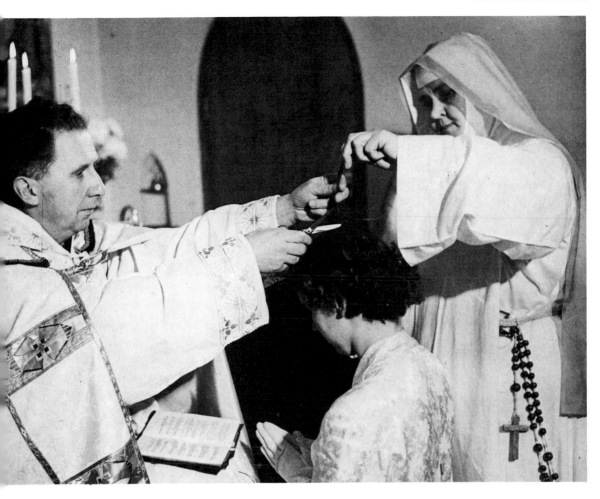

Rites of passage guarantee that important changes of role and status are made vividly memorable. A nun takes the veil.

dramatise significant changes in social status: they include baptisms, immersions, confirmations, all sorts of melodramatic initiations and humiliating ordeals, such as strippings, shavings, scarrings. Whenever we cross a threshold from one social role to another we take pains to advertise the fact with ceremonies which represent it in terms of vivid and memorable images.

The idea of 'rites of passage' was first introduced by the French anthropologist Arnold Van Gennep in 1909. Van Gennep insisted that all rituals of 'passing through' occurred in three successive phases: a rite of separation, a rite of transition and a rite of aggregation. The person whose status is to be changed has to undergo a ritual which marks his departure from the old version of himself: there has to be

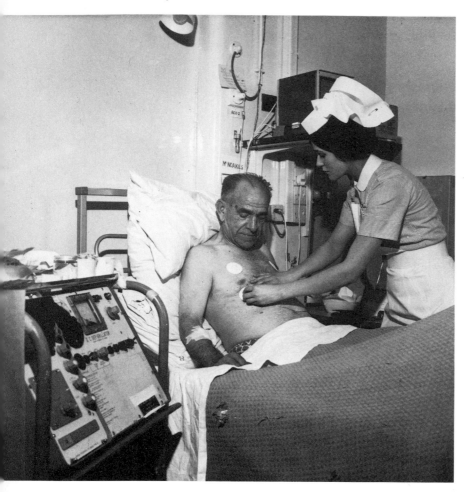

Once the person has undergone the medical 'rites of passage' he has ceased to be an agent and has become a patient — someone who suffers not only his disease but the various procedures by which it is cured.

some act which symbolises the fact that he has rid himself of all his previous associations. He is washed, rinsed, sprinkled or immersed, and, in this way, all his previous obligations and attachments are symbolically untied and even annihilated. This stage is followed by a rite of transition, when the person is neither fish nor fowl; he has left his old status behind him but has not yet assumed his new one. This liminal condition is usually marked by rituals of isolation and segregation — a period of vigil, mockery perhaps, fear and trembling. There are often elaborate rites of humiliation — scourging, insults, and darkness. Finally, in the rite of aggregation, the new status is ritually conferred: the person is admitted, enrolled, confirmed and ordained.

52

This idea can be applied to the process of becoming a patient. The fact that most of the procedures involved have a rational and practical explanation doesn't prevent them from playing a very important symbolic role as well. Although one can readily understand most of what happens to someone on entering hospital in utilitarian terms, there is no doubt that both the patient and the doctor experience some of these manoeuvres as symbolic transformations. Once someone has chosen to fall ill he has to apply for the role of patient: he auditions for the part by reciting his complaint as vividly and as convincingly as he can. This can also be seen in terms of religious confirmation: the candidate submits himself to a formal questionnaire in order to satisfy the examiner that he is a suitable person to be enrolled. If he passes the preliminary test he has to undertake the initial rites of separation. He is undressed, washed, and until quite recently, he often had to submit himself to a cleansing enema. Then come the rites of transition. No longer a person in the ordinary world, he is not yet formally accepted by his fellow patients — anxious and isolated in his novice pyjamas, he awaits the formal act of aggregation. He is introduced to the ward sister, hands over his street clothes, submits to a questionnaire by the houseman and registrar. Dressed with all the dignified credentials of a formally admitted patient, he awaits the forthcoming event.

2 · Healing and Helping

LIVING SYSTEMS HAVE A REMARKABLE CAPACITY TO re-arrange themselves in the face of any disturbance which threatens their continued existence, but if the destructive forces are severe and unremitting, they begin to deteriorate irreversibly: the various acts of adjustment interfere with one another, and the system becomes incoherent and self-destructive.

The ability to recognise and anticipate this point of no return sets man apart from other animals. A sick creature can lie low, lick its wounds and take the weight off an injured paw, but since it is not fully self-conscious, it is unable to appreciate its status as a threatened individual. Man, on the other hand, is endowed with the capacity to reclassify himself as something in need of help. Knowing that something has gone wrong, he seeks a remedy, either by using recipes which are the common property of the society in which he lives, or by consulting someone who is credited with special mastery of the healing art. Several factors determine how this credit is established. Tradition and precedent are the most obvious, though not necessarily the most important. If a healer is already known to have a large clientèle, that in itself may be enough to attract new customers, and since many illnesses get better of their own accord, and the uninformed client is incapable of distinguishing between spontaneous remissions and deliberate cures, even a quack will be able to advertise a list of satisfied sufferers. But this does not explain how a healer makes himself known in the first place.

A successful healer must identify himself with some principle which is popularly accepted as an authentic basis of medical control. He must persuade people that his methods exemplify something which they are predisposed to regard as medically effective. This applies as much to the clients of a faith-healer as it does to the patients who visit a modern cardiologist.

When an African villager submits to the regime of a local magician, he is impressed by the colourful drama with which the remedies are surrounded — by the mystery and the ritual and the esoteric confidence of the practitioner — but what sways him most is the fact that this particular type of healer acts upon the same principles of symbolic potency which the patient applies to every other aspect of his life. By the same

(On the previous pages *John Collier, 'Sentence of Death'*)

Opposite *Two forms of medical authenticity.* (Above *S. Eastman, 'A Medicine Man curing a Patient', Philadelphia Museum of Art*).

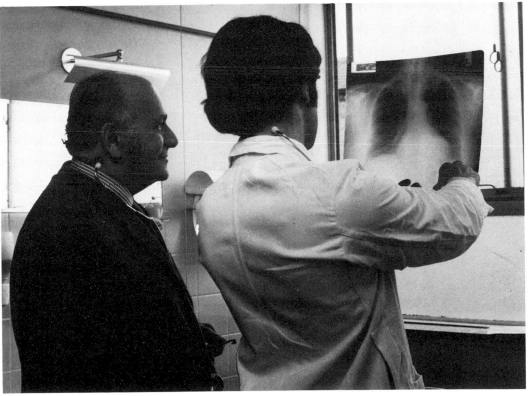

token, patients who patronise the practitioners of modern Western medicine are bolstered by the fact that such medicine is known or believed to express so-called scientific principles, which are accepted, though not necessarily intelligently appreciated, as the most orthodox mode of thought. If they turn away from conventional medicine, it is not necessarily because they are dissatisfied with its results, but because they no longer accept the rationale on which it is based.

A practitioner's effectiveness may be seen either as the expression of supernatural power — some force or energy which flows from him and produces results without the intervention of skill or technique — or as the product of special

Two forms of diagnostic curiosity. The Azande witch doctor tries to identify the human culprit responsible for his patient's illness. The Western physician seeks the physical cause.

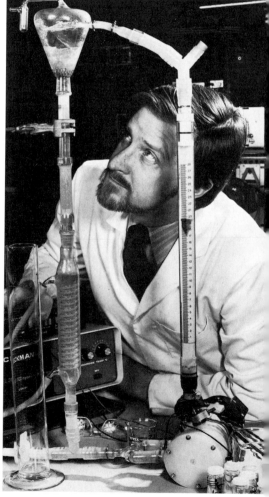

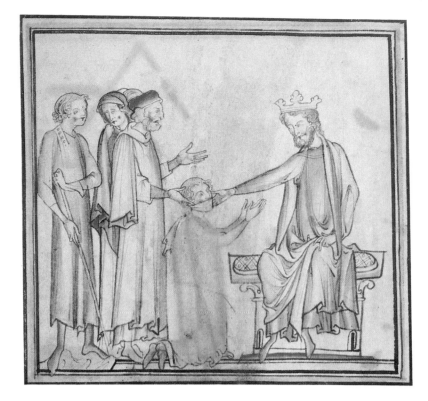

The clients of the royal healers put their faith in the supernatural power of the crown, regardless of the personal character of the king. He could be a scoundrel, although in this case he happened to be a saint. (Edward the Confessor in a thirteenth-century MS illumination, Cambridge University Library)

knowledge. Healers of the first group are for their followers the trustees and distributors of a miraculous gift, whereas members of the second group are regarded as the practitioners of a material skill. When someone chooses one or other of these two types, he is inadvertently expressing assumptions about the way in which the world works. Someone who visits a faith-healer in all probability subscribes to a belief that the world is governed by psychic powers; a patient who patronises a neurologist believes, albeit vaguely, that the world is a nexus of physical forces.

When someone is credited with supernatural medical abilities, it is often because he occupies some public office which is thought to be endowed with miraculous power. One of the most dramatic examples in the history of medicine is that of the Royal Touch. For more than 700 years, the anointed kings of France and England were regarded by their subjects as miraculous physicians whose mere touch would cure physical illness. The patient had only to come into the presence of the king and make contact with his sacred person to be healed.

Traditionally, the patients who submitted to this treatment all suffered from the scrofula, a tuberculous inflammation of the lymph glands in the neck. It has never been satisfactorily explained why this particular illness was thought to be susceptible to the Royal Touch, but the association between scrofula and monarchy had become so well-established by the Middle Ages that the illness was known as the 'King's Evil' — as if the sovereign were both cause and cure of the illness.

Sufferers ritually assembled in the royal presence, usually in a church, on the occasion of important festivals such as Easter and Ascension. In the Middle Ages these séances became a regular feature of the royal calendar, and from time to time the miraculous credit of the king was so high that it extended to inanimate objects with which his body had been in contact — the sufferers would drink the water in which the miraculous physician had washed his hands. In France the

The Royal Touch was a corporate asset of the monarchy, entrusted to the king but not personally owned by him. It was part of the official regalia.
(Left   Charles II, engraving by R. White from 'Adenochoir-adelogia', 1684.
Right   The King of France, etching from S. Faroul, 'De la Dignité des roys de France', 1633).

ritual survived until the monarchy was overthrown in 1793: Louis XVI was treating sufferers until a year or two before his execution. The same type of ceremony was inaugurated in England by Henry I, and the ritual did not die out until the accession of the Hanoverians.

The association between monarchs and miracles has a long history. Amongst the German tribes of pagan Europe the king or chieftain was widely credited with being able to intervene in the course of nature, to turn aside the threats of flood, famine and pestilence. However, these powers remained vague until the installation of Christian monarchy, and the ceremony of the Royal Touch was not formally inaugurated until European kings were installed by the rite of coronation, which, by re-enacting the baptism of the body, identified the monarch with the supernatural authority of Jesus Christ. Ironically, the rite which helped to confirm the Royal Touch transferred it from the person of the king to the office of monarchy, and his medical power was not allowed to come into play until he had been anointed and crowned. The miracle was automatically transferred from one monarch to the next, and although the mortal remains of the dead king

Left *The title page of 'Adenochoiradelogia'.*
Right *Constitutional monarchs make no claims to supernatural power. But even in a modern democracy loyal sufferers derive comfort from the smiles of an anointed sovereign.*

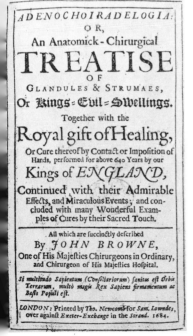

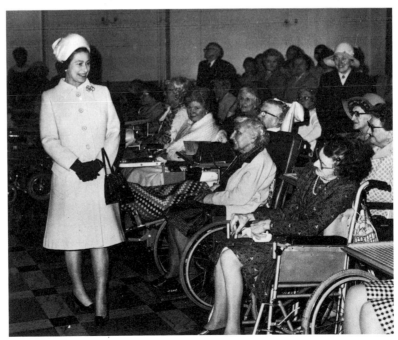

The inauguration of Christian monarchy created a theocratic type of royalty. Anointed by priests, the king became God's deputy, entrusted with supernatural authority over his subjects.
(Barend van Orley, 'Consecration of the King of France', Galleria Sabauda, Turin)

continued to inspire awe and respect, they were, unlike the relics of a saint, medically inert.

This expressed a fundamental change in the attitude towards the monarchy itself. Before the establishment of Christianity, the legitimacy of the king was a matter of his personal strength, cunning or virtue, although it was usually reinforced by his descent from a royal line. During the Middle Ages, however, constitutional theorists distinguished between the mortal person of the king and the immortal office he occupied. The king, it was said, had two bodies, one natural and one politic:

> His Body natural (if it be considered in itself) is a Body mortal, subject to all Infirmities that come by Nature or Accident, to the Imbecility of Infancy or old Age, and to the like Defects that happen to the natural Bodies of other people. But his Body politic is a Body that cannot be seen or handled, consisting of Policy and Government, and constituted for the Direction of the People, and the Management of the public weal, and this Body is utterly

Opposite Baptism conveys and confirms the gift of supernatural grace. Healing transmits the grace to ordinary mortals.
(Verrocchio/Leonardo, 'Baptism of Christ', Uffizi, Florence)

void of Infancy, and old Age, and other natural Defects
and Imbecilities, which the Body natural is subject to,
and for this Cause, what the King does in his Body
politic, cannot be invalidated or frustrated by any
Disability in his natural Body ...

   This distinction is vividly expressed in the funeral effigies
which are to be found in the royal mausoleum at St Denis
near Paris. The tombs contain not one but two statues of the
king: a naked figure representing the mortal remains of the
man himself, and a gorgeously robed effigy which stands for
the imperishable office of the monarchy. Nowadays, we regard the naked statue as a reminder of the fact that even kings
must die. But for the men of the Middle Ages the clothed
figure made the much more important point that the institution of monarchy was indestructible. When the divinity
of the office was called into question, the medical miracle
associated with it suffered a severe setback. Although
Charles II reinstated the practice following his restoration,
the ceremony never fully recovered its former vigour, and by
the time Samuel Johnson was brought up to London to
receive the touch of Queen Anne, it was little more than a
quaint survival.
   Miraculous healing is not necessarily related to high office.
Sufferers can just as easily acknowledge the supernatural
claims of a private individual. Jesus of Nazareth deliberately
turned his back on rank and title. The fact that his miraculous
powers were associated with his personality means that he
belongs to the group of persons whom the German sociologist
Max Weber declared had 'charisma', by which he meant not
personal glamour but something which might be regarded as a
sign of heroic endowment: superhuman strength, holiness,
chastity or zeal:

   The charismatic hero does not deduce his authority from
   codes or statues, nor from traditional customs, nor from
   feudal vows of faith. He gains and maintains his authority
   by proving his strength in life.

   Weber was describing the sociology of great leaders, and
although some of the figures he discussed used faith-healing
as a way of advertising or clinching their missionary claims,

*The tomb of François I in the
royal mausoleum at St Denis.
The immortal body of the
monarch is represented by the
praying figure on top; the
mortal individual awaits
resurrection below.*

64

Weber did not concern himself with the medical details. Nevertheless, the idea of charisma helps to distinguish those healers whose power depends on personal character from the ones whose remedies are invested in sacred rank. The scrofulous patients who crowded into the royal presence put their faith in the office of the Crown; the woman who reached her hand to touch the hem of Jesus's garment recognised the personal appeal of the Saviour himself:

> And a woman having an issue of blood twelve years, which had spent all her living upon physicians, neither could be healed of any, came behind him, and touched the border of his garment: and immediately her issue of blood stanched. And Jesus said, Who touched me? When all denied, Peter and they that were with him said, Master, the multitude throng thee and press thee, and sayest thou, Who touched me? And Jesus said, Somebody hath touched me: for I perceive that virtue is gone out of me. And when the woman saw that she was not hid, she came trembling, and falling down before him, she declared unto him before all the people for what cause she had touched him, and how she was healed immediately. And he said unto her, Daughter, be of good comfort: Thy faith hath made thee whole; go in peace [Luke IX, 43–8].

The miraculous authority of the prophet Elijah was also personal in character, although the miracles were performed through rather than by him; whereas Jesus Christ, as the incarnation of God and not merely his earthly representative, was presumably the author of his own remedies. In both cases, though, the therapeutic power was invested in the living person of the miracle worker himself.

Neither Jesus nor Elijah offered his cures as a routine out-patient service. They used their powers somewhat sparingly, and when they did stoop to heal the sick it was with the ulterior motive of dramatising their mission and making a point about the miraculous power of faith. This is almost the reverse of modern faith-healing, in which the healing is an end in itself. Nowadays faith-healers hold regular clinics, and, although these are usually cast in the form of religious services, which means that the cures are regarded as sacraments and not simply treatment, the congregation has a particular

Above   *A charismatic healer may become associated with one particular form of disease, sometimes because he himself was afflicted with it and then was miraculously relieved of the illness in question. St Roche thus became traditionally associated with the plague. (Quentin Massys, detail of altarpiece, Alte Pinakothek, Munich)*

Opposite   *As with the Royal Touch the cures performed by charismatic healers are accomplished by an unmediated fiat. No skill or technique intervenes. ('Raising the Dead' and 'Curing the Paralytic', sixth-century Byzantine mosaics from Sant Apollinare Nuovo, Ravenna)*

interest in dramatic clinical improvements.

The gift of miraculous healing can just as easily be recognised in someone who claims that and nothing more —
especially when affliction is widespread and unrelieved. His
charisma may, in fact, consist of nothing more dramatic than a
convincing account of the moment when he was called to
exercise it. One of the most remarkable instances of this is the
case of the seventeenth-century Irish faith-healer Valentine
Greatrakes. Greatrakes was an ordinary country squire, and
until he was in his early thirties, he showed no signs of being
marked out for a miraculous career. But, shortly after the
Restoration of Charles II, he felt an irresistible urge to practise faith-healing:

> About four years since, I had an impulse or a strange
> persuasion in my own mind (of which I am not able to

*Valentine Greatrakes, the
seventeenth-century Irish
healer.*

The touch of a faith healer is not in any sense a handicraft. As with Jesus the virtue flows from him by courtesy of the supernatural power which employs him as its passive instrument.

give any rational account to another) which did very frequently suggest to me that there was bestowed upon me the gift of curing the King's Evil, which for the extraordinariness of it, I thought fit to conceal for some time but at length I communicated this to my wife, and told her, I did verily believe that God had given me the blessing of curing the King's Evil; for whether I were in private or public, sleeping or waking, still I had the same impulse ...

During the Civil War Greatrakes had served as an officer in Cromwell's army, and the fact that he was prompted by this urge so soon after the Restoration leads one to suspect that there was a thinly veiled Republican motive, that he was eager to prove that healing powers were not restricted to the monarchy. His first séance proved amazingly successful, and before long he was touching and stroking a rabble of sufferers from all over Ireland — not only for the King's Evil but for many other illnesses as well. His reputation soon reached England, and in 1665 he was invited to cross the water and

bring relief to a somewhat more distinguished patient, Viscountess Conway. Ann Finch was one of the most remarkable women of her age. As a girl she taught herself Latin, Greek and theology; she mastered astronomy, alchemy and some of the more abstruse traditions of mystical philosophy. These achievements would have been remarkable in a healthy man, but in a seventeenth-century woman who suffered for nearly thirty years from blinding migraine they were little short of miraculous. She is the first in a line of invalid blue-stockings —

In the healing of Naaman the power seems to flow directly from the hand of God, as if the curing of Man was a re-enactment of the process by which he was first created. (Champleve enamel panel, c. 1180, British Museum)

clever women who seemed to organise their intelligence through the medium of their crippling illnesses. It would be uncharitable to imply that such writers as Elizabeth Barrett Browning and Harriet Martineau consciously used their illnesses to manipulate the loyalty and affection of their friends, but there is no doubt that the concern which their symptoms created helped to consolidate the intellectual circle on which they depended. Lady Conway had been treated, advised, consoled and inspired by some of the most sophisticated minds in England, including William Harvey, who at one time had acted as her personal physician. Her constant companion was the great Neo-Platonist philosopher Henry More, Isaac Newton's friend and teacher.

When news of Greatrakes's cures reached England, Lady Conway's husband persuaded the Irish healer to visit them on

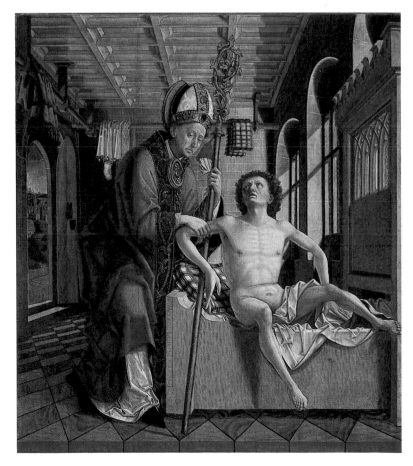

The charismatic healer is typically someone without high office or impressive status. But personal saintliness is not incompatible with robes and vestments and may even become identified with them.
(Michael Pacher,
'St Wolfgang healing a Sick Man', detail of altarpiece, Alte Pinakothek, Munich)

their family estate. He had a successful session with the local tenants and was then allowed to lay hands on the aristocratic invalid herself. But the famous migraines persisted, and Greatrakes made an excuse to leave. By this time he had become the centre of a national controversy. Some thought he was a saint, others were convinced he was a charlatan. More serious observers tried to explain his successes in 'scientific' terms:

> I have seen pains strangely fly before his hand, till he hath chased them out of the body, Dimness cleared and Deafness cured by his touch; twenty persons at several times, in fits of the Falling-Sickness, were, in two or three minutes, brought to themselves, so as to tell some extreme part ... But yet I have many reasons to persuade me, that nothing of all this is miraculous. He pretends not to give Testimony to any Doctrine, the manner of his Operation speaks it to be natural, the Cure seldom succeeds without reiterated touches, his Patients often relapse, he fails frequently, he can do nothing when there is any Decay in Nature, and many Distempers are not at all obedient to his touch: so that I confess, I refer all his virtue to his particular Temper and Complexion, and I take his spirits to be a kind of Elixir and Universal Ferment.

In all probability the writer was using the term 'Elixir' and 'Universal Ferment' to refer to the ancient mystical notion of the Aether — a spiritual substance half-way between mind and matter, traditionally thought to be capable of transmitting influences from one end of the universe to the other. This notion runs like an underground stream in European thought from the Greek Stoics onwards. It played an important part in Plato's thought, and whenever his philosophical principles were revived, the idea that the world was animated by a cosmic spirit reappeared. Ironically the members of the Conway circle regarded themselves as inheritors of Plato's mysticism. Lady Conway, for instance, thought that:

> all Creatures from the highest to the lowest are inseparably united one with another, by means of Subtiler Parts interceding or coming in between, which are the Emanations of one Creature into another, by which also

The notion of an inter-penetrating spirit runs throughout European history. In this arcane diagram by the Renaissance magician Robert Fludd it mediates between the mind of God and the body of the physical universe.

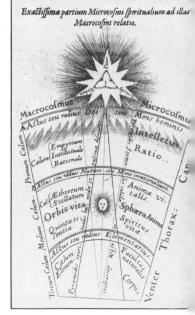

they act one upon another at the greatest distance; and this is the Foundation of all Sympathy and Antipathy which happens in Creatures: And if these things be well understood of any one, he may easily see into the most secret and hidden Causes of Things, which ignorant Men call occult Qualities.

Franz Anton Mesmer. He established an apostolic succession of magnetic healers. Amongst his French and English followers his theory of a 'universal fluid' helped to create a tradition of millennial brotherhood, a Utopian future in which all social discord would dissolve.

But though the Conways probably regarded Greatrakes as a living proof of this theory, it is unlikely that the man himself knew or cared much about Plato's cosmic spirit. As far as one can tell, he saw himself as someone who had received the unsolicited gift of God's grace. If he had cured Lady Conway, he might have been tempted to rationalise his gift in Neo-Platonist terms. But his mission faded away before it had time to acquire a theoretical justification.

The fact that a mission is based on charisma does not preclude the use of a justifying theory. A practitioner may start his career with an interest in some rational principle and go on to mount a healing mission in order to prove his point, acquiring charisma on the way. This is more or less what happened to the famous hypnotist Franz Anton Mesmer.

Mesmer was born in 1734. He studied medicine in Vienna and in 1766 published what would strike most modern readers as a somewhat eccentric thesis on the influence of the stars and planets on human health. In some respects the essay was a throwback to traditional astrology, but, as a product of the eighteenth-century Enlightenment, Mesmer tried his best to reconcile the occult superstitions of the Renaissance with the scientific principles of the Age of Reason. The English physicist Isaac Newton provided him with a convenient rationale. At the end of his *Principles of Natural Philosophy* Newton had invited his readers to consider the possibility that the universe was pervaded by an Aether, and that this might explain the transmission of light, magnetism and gravity. Although Newton put this forward only as a tentative afterthought to his famous theory, it was seized upon by those who needed scientific justification for their belief that the universe was animated by occult forces.

After some abortive experiments designed to show the influence of metallic magnets on the human body, Mesmer decided that the unaided mind was capable of transmitting

73

*Opposite   Ann Finch,
Viscountess Conway, was the
centre of a Neo-Platonist circle
whose members believed that
the universe was animated by a
spirit which rolled through all
things.
(Samuel van Hoogstraaten,
Mauritshuis, The Hague)*

influences from one person to another, a transmission he
called 'Animal Magnetism'. In one of his later 'Queries' in
Optics Newton had put forward the suggestion that human
nerves were filled with aether, and that by travelling through
this imponderable medium the motives of the mind could
bring about the movements of the muscles. If it was true that
the nerves were filled with the fluid that pervaded the universe, it was quite reasonable to assume that the mind could
broadcast its energy across space. Although his revised theory
did not require the use of magnets, Mesmer continued to
employ metallic apparatus to demonstrate the physiological
effects of the universal force. He invited his patients to sit
round a wooden tub filled with iron filings — the container was
presumably supposed to act as an accumulator. When the
sufferers gripped steel bars inserted into the tub, they obedi-
ently convulsed and fell into trances, after which, according
to some of the more favourable reports, they were relieved of
their symptoms. Arthritis, eczema, headaches and hysteria —
diseases which are now recognised as having a large psycho-
somatic component — proved particularly susceptible.

When Mesmer started his 'magnetic' clinic, he immediately encountered the envy and suspicion of the Austrian
medical establishment, although it is difficult to know by what

*Sceptical cartoonists took a
somewhat less than reverent
view of Mesmer's magnetic
claims.*

MAGNETIC DISPENSARY.

criteria the profession distinguished Mesmer's remedies as examples of quackery. 'Magnetism' was no more irrational, no more unscientific, than the orthodox remedies of purging, blood-letting and cupping. Even at the end of the eighteenth century, so-called conventional medicine was a tissue of contradictions: there were no consistent intellectual standards and no organised body of scientific principles. Nevertheless, Mesmer's opponents probably recognised that, for all its Newtonian credentials, his magnetic remedy was essentially mystical in character and therefore revived the monkish superstition of the Middle Ages. Whatever the reason, Mesmer was forced to emigrate to France, where he soon established a successful salon in the Place Vendôme. By this time he had acquired the demeanour of a charismatic healer: he stalked the darkened salon in a long robe embroidered with astrological symbols and took care to accompany the séances with entrancing music. Everything was done to produce a susceptible state of mind among the fashionable clients.

Paris was soon seized by a Mesmeric mania, and in 1784 a French Royal Commission was set up to investigate the claim. The committee, which included both Benjamin Franklin and the great chemist Lavoisier, almost unanimously agreed that the results, such as they were, could all be explained as the outcome of suggestion and that there was no need to invoke the agency of a cosmic fluid.

The conclusions of the committee expressed a growing recognition of the powerful hold imagination has over the body. Suggestion was used to account for traditional remedies like the Royal Touch and for new-fangled cures such as the ones offered by Mesmer. In 1800 it was invoked to explain yet another medical fad. An American called Perkin had introduced the use of metal tongs, which he called 'tractors'. He claimed that the magnetic field created by these would draw out the evil in an arthritic limb, leaving it mobile and painless. The West Country physician John Haygarth was convinced that this was yet another example of the imagination at work and, to prove his point, he made a pair of wooden tongs and painted them with black lead to give them a metallic sheen. As he predicted, the credulous patient was usually relieved of his pain. Like Mesmer, Haygarth appreciated the influence of the physician's demeanour: he emphasised the vital importance of

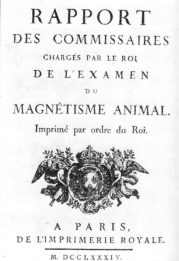

RAPPORT
DES COMMISSAIRES
CHARGÉS PAR LE ROI
DE L'EXAMEN
DU
MAGNÉTISME ANIMAL.
Imprimé par ordre du Roi.

A PARIS,
DE L'IMPRIMERIE ROYALE.
M. DCCLXXXIV.

Following the report of the Royal Commission in 1784 Mesmer's personal credit was virtually extinguished. The magnetic craze outlived its creator by more than fifty years and found enthusiastic disciples in England, especially during the 1840s.

keeping a straight face, and brandished a stop-watch as if to show the patient that serious scientific action was afoot.

Perkin's 'tractors' were soon consigned to the obscurity they deserved, but although Mesmer suffered a personal eclipse following the Report of the Royal Commission, Animal Magnetism outlasted its inventor by more than fifty years. It continued to flourish well into the nineteenth century. One of the most interesting episodes took place in London in 1838. John Elliotson was a professor of medicine at the newly established University College Hospital in London. As a practising physician and a medical author, he was widely acknowledged as a distinguished leader of the profession: he had published an influential medical textbook and was personally responsible for the introduction of the stethoscope into England. He was a close personal friend of Charles Dickens and attended both Thackeray and Wilkie Collins. *Pendennis* is dedicated to him, and he actually appears in the novel, as Dr Goodenough. Until 1836 Elliotson showed no particular interest in Animal Magnetism, although he had witnessed a demonstration of it nearly ten years before. In 1837, however, he underwent a dramatic conversion as a result of meeting one of Mesmer's apostolic successors, the Baron Dupotet. Almost overnight, Elliotson became a convinced Mesmerist, and with the Baron's help he instituted the Mesmeric treatment of his hospital patients. Fortunately, his case-books have been preserved, so that it is possible to trace the course of his new career.

The most dramatic case involved two sisters, Elizabeth and Jane Okey, who had been admitted to the hospital suffering from fits — it is hard to tell from the reports whether these were hysterical or epileptic. The girls were put into a magnetic trance and appeared to lose their symptoms. The improvement was short-lived, however, and the treatment had to be repeated several times a day. Soon patients and physician were locked in a self-perpetuating cycle of complaint and treatment. Not that there was any conspiracy or collusion, but there is no doubt that Elliotson and the Okeys unconsciously satisfied complementary needs, and the notion of permanent cure inevitably gave way to one of interminable treatment. The hysterical sisters flourished under the special attention which their behaviour in trance excited, and it was obviously to Elliotson's advantage to keep a hold on his colourful patients. Doctor and patients became an inseparable trio — a three-headed monster of magnetic exorbitance. Before long, the two sisters began to exhibit clairvoyant powers. By the end of 1838 Elliotson claimed that they were able to visualise their own innards, and as he deepened their trance the clairvoyance extended to the interiors of other people. Soon Elliotson was steering them around the wards, using them to diagnose the state of other patients' organs.

The Okey girls, it seems, were not altogether unfamiliar with visionary experiences. Before their admission to hospital they had been members of Edward Irving's Catholic Apostolic Church in Regent Square in Bloomsbury. The services of this sect were marked by ecstatic interventions in which the witness spoke in foreign languages — so-called 'glossolalia'. It is clear that the two girls had unconsciously learnt to modulate their fits according to the social context in which they occurred: in church the sisters were regarded as pentecostal

John Elliotson, 1791–1868. Professor of Medicine at University College, London, he was a brilliant clinician with an inventive scientific curiosity. He introduced the stethoscope into English medicine and translated Blumenbach's 'Physiology'. He was a friend of Dickens, Thackeray and Wilkie Collins. His enthusiasm for Mesmerism brought him into conflict with his colleagues and in 1838 he resigned from his professorship. He founded and edited 'Zoist', a journal of phreno-mesmerism.

prophets, and in hospital they were accepted as Mesmeric shamans. The conclusion was inevitable. Quite understandably, Elliotson's conduct outraged his hospital colleagues. At that time the medical profession was almost within reach of coherent scientific credibility, and demonstrations of this sort threatened to destroy all that had been won. He was asked to discharge the Okey sisters, and, to avoid further humiliation, he resigned.

But Mesmerism continued to excite considerable controversy, and for nearly twenty years Elliotson edited a successful journal called *The Zoist*, which was largely devoted to the description of magnetic successes. By the mid-1840s, however, the subject had begun to undergo a fundamental change, largely through the work of a Manchester surgeon called James Braid. In 1843 Braid published an influential book, *Neurypnology*, which developed the sceptical conclusions of the French Commission. Like his French predecessors, Braid refused to acknowledge the existence of Mesmer's magnetic fluid and attributed the trances to the psychological suggestion exerted by the mesmerist. He also emphasised the importance of fixing the subject's gaze, insisting that it was the enforced monotony which this brought about that caused the subject to lapse into a reverie. He even renamed the phenomenon, calling it 'nervous sleep' or 'hypnotism'. Braid's doctrine found enthusiastic support, and the French physicians who exploited his ideas in the 1860s and 1870s laid the foundation for the psychoanalytic use of hypnotism. Mesmer's theory led indirectly, but none the less inexorably, to the revolutionary work of Freud.

By the end of the nineteenth century most of the remedies discussed in this chapter would have been officially dismissed as the results of suggestion, which was a way of distinguishing them from cures based on physiological principles. Today, doctors have a more sophisticated interest in such phenomena and are aware that the effectiveness of scientific medicine depends to an almost embarrassing extent on the patient's recognition of both the official status and the personal demeanour of the doctor. The concept of suggestion is now enshrined in something which is widely known as 'the patient–doctor relationship'. This implies that what takes

The healing power of the physician depends to an embarrassing extent on the patient's belief and confidence. This confidence may not be objectively justified. This physician, for instance, had almost nothing realistic to offer but in the absence of anything better the patient was at least consoled by his attention. (Gerits van Brekeleunkam, 'The Consultation', Louvre, Paris)

place between a sick person and the doctor who treats him is an elaborate two-way transaction, the outcome of which is determined not simply by what the doctor hands out, but by the confidence and respect the patient gives him in return. It would be an exaggeration, of course, to equate a doctor's rank and qualifications with the mystical aura which surrounded a medieval king — the physician doesn't claim, and the patient certainly doesn't recognise, supernatural authority. Nevertheless, the doctor's importance and even title often has a bearing on the patient's recovery. Hospital patients often flush with excitement at the expected visit of an important specialist, and since the Grand Round is usually marked by all sorts of impressive panoply including a respectful retinue of residents and students, the ward is put into a state of receptive reverence, and even terminal cases may declare that they feel much better. Not a Royal Touch, perhaps, but a good secular substitute.

Only the most narcissistic doctor would dignify his personal appeal with the name of 'charisma', but although his peculiarities are not interpreted as signs of miraculous endowment, it is an inescapable fact that glamour, good looks, verve and confidence can all help to make the physical treatment more effective. In fact, a doctor's optimism can be so infectious that it can seriously interfere with the accurate appraisal of a new treatment. Until comparatively recently, new drugs were tested by dividing patients into two groups: one group received the active pill; the other was dosed with an inert blank. When drugs such as cortisone were tested in this way, they gave dramatically favourable results until it was realised that the doctor's own behaviour had an important part to play in the outcome. When giving the supposedly active pill the doctor betrayed his optimism to the patient, who then overresponded accordingly. Now, patient and doctor are both kept in ignorance about the pills which are being used. When this technique was applied to one 'miracle drug', it was found that its reputed effects were partly due to the fact that it was the doctor who was expecting miracles.

In the case of faith-healing, both remedies which are attributed to a sacred office and those which are credited to a remarkable individual share one significant feature: they do not involve any voluntary effort. No doubt the king was exhausted at the end of a hard day's touching, and even Jesus spoke of the virtue having flowed out of him, but neither the monarch nor the Messiah had to exert himself — at least not in the ordinary sense of the word. Although we commonly speak of someone 'performing' miracles, a miracle is not performed in the way that handstands, concertos or surgical operations are. You can't perform a miracle badly, and you can't improve your record by practising or rehearsing. If it were possible to perfect a miracle by practising it, one would be reluctant to regard it as a miracle in the first place; one would tend to call it an accomplishment. Miracles are not even actions, or at least not voluntary ones. They are, in a sense, out of the miracle-worker's control, though this doesn't mean that they are involuntary like sneezing, shivering or blushing — people do not have fits or spasms of miracle-working.

The right hand of the Creator touches Adam into life. (Michelangelo, Sistine Chapel ceiling)

Pottery is a typical way of making things happen. There is an intelligible relationship between what is done and how things turn out. The movements of the hand visibly shape the outcome.

The role of the person who performs miracles is consequently extremely obscure: his presence is essential, but his effort is irrelevant. Mesmerism is ambiguous in just the same way. Monarchs, messiahs and mesmerists all broadcast an unmediated influence — unmediated because nothing resembling technique or dexterity is involved. The Royal Touch never degenerated into a massage, and although Valentine Greatrakes was known as 'the Irish stroker', he was not simply a primitive osteopath. The manual gestures used, like the 'passes' made by a mesmerist, may have been indispensable features of their performance, but they did not visibly shape the outcome in the way that a potter's hand shapes what he is making. In spite of the emphasis which they gave to the hand, one cannot regard any of the monarchs, messiahs or mesmerists as practising manual skills.

Medicine of this sort exemplifies grace as opposed to works. As long as the patient acknowledges that he needs help, he has only to present himself in a suitable state of submission and the supernatural influence will do the rest. In such a context, disease is analogous if not identical to sin: the sufferer is redeemed and renewed by what pours forth from the baptising hand of the healer.

Such healers continue to play a significant part in the history of medicine, but they are the exception rather than the rule. Most remedies, whether they are effective or not, represent applied knowledge and involve the use of physical procedures: recipes, incisions or manipulations, which are dictated by a body of ideas which have to be acquired and understood by the person who acts upon them.

The medical wisdom of one community is not necessarily acknowledged by members of another. Western physicians regard the magic and sorcery of tribal Africa as mumbo-jumbo, but, as far as the tribal practitioner is concerned, the spells, incantations and brews express a system of ideas in which there is an intelligible relationship between what he does and what he hopes to bring about. And, as in Western medicine, this presupposes a diagnostic curiosity. Unlike the faith-healer, whose cures are believed to be comprehensively beneficial, the expert healer, savage or civilised, must identify what he is dealing with: he must size up the situation, distinguish one illness from another and recognise what its

susceptibilities are likely to be. He looks for causes and tries to treat them.

But before looking for causes he must have a list of relevant possibilities in mind. All medical enquiries take place within a more or less circumscribed domain of theoretical commitment in which curiosity is limited by what the investigator regards as possible. Within the range of what is possible he arranges alternatives in order of their probability. Investigation presupposes suspicion. The suspicion may be very generalised, but even at its vaguest it prompts the investigator to look in one direction rather than another. For example, a few years ago, when a number of American legionnaires attending a conference in Philadelphia unaccountably fell ill and died, the public health authorities started their enquiry by looking for poisons in the food and in the air-ducts of the hotels. When that proved fruitless, they searched for, and

For the Azande illness and misfortune may be the result either of witchcraft or sorcery. The witch takes no voluntary action to inflict his mischief whereas the sorcerer must deliberately brew magic spells.

eventually found, a virulent organism.

It was more than a year before the investigation was successfully concluded, but although the first suspicions were extremely vague the investigators were never in doubt about the range of possibilities: it had to be either a toxin or a living organism. Such alternatives are implied by one of the most basic assumptions in Western medicine; namely, that all serious disease has an identifiable physical cause. If the search for organisms had proved as fruitless as the one which had previously been mounted for poisons, the investigation would still have been directed towards a physical agency, and even if the case had remained open for another twenty years it is very unlikely that psychic causes would have been considered — even in desperation.

For a tribal African, both the possibilities and the probabilities would have been altogether different, not simply because toxins and bacteria are outside his range of experience, but because his domain of suspicion would not include physical agents. For him, disease is the domain of psychic rather than physical causes, and the wrath of spirits or the malevolence of neighbours would have been the plausible alternatives. But his investigation and treatment would involve just as much expertise as the American doctor's.

The varieties of tribal healing are so numerous that one could devote a whole book to them alone. I have chosen one particular example not because it is more important than all the others or even typical, but because it illustrates a form of medicine in which the diagnostic interest is almost the opposite of our own. When in the late 1920s the British anthropologist E. E. Evans-Pritchard went to live among the

The anthropologist E. Evans-Pritchard amongst the Azande.

86

Azande of the southern Sudan, he was almost immediately struck by their distinctive attitude to physical illness. Mild or self-limiting conditions were usually treated on an *ad hoc* basis, but whenever members of the tribe felt seriously ill they almost invariably attributed it to the malevolence of friends, relatives or neighbours.

There were, Evans-Pritchard discovered, two types of malignant influence: witchcraft and sorcery. In Western Europe these two are often confused, but the Azande see them as quite distinct, and the distinction corresponds to the one I have already drawn between faith-healers and expert physicians.

The witch, who can be male or female, is said to be born with his or her dangerous endowment. This is the so-called 'witch-organ': a large cyst or swelling which can be found at post-mortem, usually attached to the underside of the liver — or so they say. By the time Evans-Pritchard arrived in the Sudan, the practice of conducting post-mortems on suspects had fallen into disuse, partly because it was disapproved of by the colonial administrators, and when I visited the tribe in 1977 no one was able to remember when the last autopsy had been conducted. Nevertheless, belief in the existence of witches persists. The witch apparently has no voluntary control over the malignant effects of the organ. Anger or grievance, emotions which are supposed to overheat the organ, may exacerbate its activity, but since the individual has little or no control over his feelings, and since the witch organ is inherited, the witch is the innocent victim of his or her dangerous endowment. Indeed, the person who is accused of it generally feels dismayed rather than guilty. It is this which distinguishes a witch from the other type of mischief-maker. A sorcerer is the conscious manipulator of bad magic, a spiteful technician, who brews or concocts his wicked mixtures according to a recipe, reinforcing their effectiveness with deliberate spells or incantations.

When a Zande tribesman comes down with an illness which cannot be readily accounted for, little or no attempt is made to establish the identity of the illness: if the pain, discomfort or inconvenience is serious or persistent enough, he automatically consults an oracle to find out who is harming him.

There are four oracles, all of which involve the use of some physical material which can exist in one of two states. A question is posed, and the answer is provided by the way in which the material behaves. The simplest oracle is the Mapingo. Three short sticks of equal length are placed in a neat pile, two on the bottom, one on the top. The question is then put in the following manner: 'If my illness is caused by the witchcraft of so-and-so, Mapingo scatter.' The heap is left unsupervised for twenty-four hours: if the sticks are disturbed, the suspicion is confirmed. The rubbing-board, or Iwe, works on a similar principle. After moistening the surface of a small, polished tripod, the operator begins to rub it with another piece of polished wood. While he rubs, he murmurs, 'If it is so-and-so who has caused my illness, Iwe, stick after three rubs', and if the board jams on the third rub, the 'diagnosis' is confirmed. The so-called 'termite oracle', considered to be more authoritative than either of these, involves two twigs, one sweet and one bitter, which are pushed together into the hard earth of a termite nest while the operator invites the termite to nibble one or other. After twenty-four hours, the sticks are withdrawn and their ends are examined.

If these three oracles give inconclusive answers, the patient consults the most important oracle of all — the Benga, or poison oracle. A chicken is forcibly fed with the extract of a poisonous vine, and the gagging bird is commanded to die if the person named is the culprit and to survive if he is blameless. Unlike the other three oracles, the Benga is surrounded by a supernatural aura, and its use is attended by elaborate rituals and precautions. The poison can only be obtained from plants which grow a long way away, and in Evans-Pritchard's day, when transport was bad and the bush was much thicker than it is now, this involved a long and often quite dangerous journey. The person who fetched the poison was expected to refrain from sexual intercourse during the expedition and to abstain from certain articles of diet. The oracle could only be consulted in great secrecy and the operator was obliged to withdraw to the edge of the woods. Those who approached him did so with fear and reverence, treating him with the sort of respect which was normally given only to persons of royal blood. In my first interviews with the Azande I found them reluctant to admit the use of this oracle: they said that it was an old superstition and that no Christian Zande would now consult it. However, on closer acquaintance, it became clear that the oracle was more frequently consulted than they had cared to say — and several old men were prepared to demonstrate it.

As a last and somewhat more expensive resort, the victim may consult a human expert: a witch doctor. Dressed in a traditional skirt of monkey skins, wearing a straw hat topped with parrot plumes, the witch doctor performs a diagnostic dance. He works himself up into a frenzy, assisted by the drumming and chanting of interested villagers. At the height of his excitement, he comes to a sudden standstill, holds out his hands for silence, and announces his decision. In Evans-Pritchard's day this was usually followed by a short trip to the hut of the suspect, where a chicken wing was accusingly flung down at the culprit's feet. The accused would either deny responsibility or, if he was convinced by the authority of the witch-doctor, try to deflect vengeance by attempting to 'cool' his witchcraft, usually by taking a mouthful of water and blowing it out, followed by the announcement that he was trying to the best of his ability to lower the temperature of his witch-organ.

As in Western medicine, the questions which are put both to the oracle and to the witch doctor express what the patient thinks is the possible cause of his disease. But since the oracles can provide only yes/no answers, it is no good asking them who is responsible: you have to have quite specific suggestions. When someone falls ill in Zandeland, he does not merely suspect witchcraft in general; he attributes his illness to the witchcraft or sorcery of particular people with whom he already has a quarrel. In other words, the diagnostic techniques employed by the Azande articulate pre-existing grudges and grievances.

To a large extent, Zande medicine is an extension of tribal law or morality: illness is analogous to our notion of tort — at least as far as its diagnosis is concerned — and the role of patient more or less parallels that of plaintiff. When a member of this tribe falls seriously ill he is seeking not only relief, but vengeance or compensation, and the action frequently ends up in court, where the culprit is usually advised to refrain from his or her spiteful practices and/or to pay compensation for the damage that has been done. Naturally, this is rein-

Azande witch doctor about to perform his diagnostic dance. For him and for his clients, the daubs of dust and skirt of monkey skins are essential parts of his professional equipment.

Opposite
The poison oracle and the rubbing board.

forced by magical measures to counteract or turn aside the dangerous influence.

In the West the notion of the witch or sorcerer is traditionally associated with fear and dread, darkness and horror, and has given rise to hideous excesses of sadistic revenge. Evans-Pritchard emphasised that these overtones were altogether missing among the Azande, who regarded witchcraft and sorcery as one of the chronic facts of life, like gossip and scandal. In later years when his students asked him what it was like to live among people who held beliefs of this sort, he said that it was only too easy to identify with them. In the first few weeks of his stay, he had been startled and even shocked by the way in which these ideas conflicted with Western notions of cause and effect, but since the concept of magic permeated every aspect of their daily life — from the making

of pots to the planning of hunting expeditions — it was impossible to enjoy any intimacy with them without yielding to their system of thought, and after a few months he found that he, too, was suspecting human malice whenever he fell ill or injured himself.

However, endemic diseases such as malaria or sleeping sickness are regarded by the Azande not as the result of witchcraft but as one of the natural risks of life, and if they think about their causes at all, they attribute them to the actions of the Supreme Being — someone or something which they call 'Mbori'. When I visited the tribe I noticed a surprisingly high incidence of goitre — almost one in every twenty women seemed to be afflicted with this large, soft swelling of the thyroid. But when I asked whether this was the result of *mangu* or witchcraft, they insisted that it was one of those things, like hernia in men, that just happened. Oracles or witch doctors are consulted only when the ailment is severe enough or persistent enough to arouse suspicion of a human cause. In cases of mild illness or trivial injury, the situation is usually treated at its face value: an injured foot is bound and poulticed; an abscess will be cleaned and often scalded with hot water.

The Azande make an intuitive distinction between illnesses where the patient feels generally awful and those which merely affect local parts. Fevers, headaches and abdominal pains are almost invariably treated by remedies taken by mouth. But when the disorder is confined to a limb or to the skin, the patient simply dabs, binds or anoints the affected part. The Azande share a worldwide conviction that the human physique is divided into two distinct sectors: a metropolitan self with which personal identity is associated, and a series of outlying protectorates whose disordered function may produce suffering without threatening the integrity of the individual. Our own language expresses this distinction very vividly. A patient with a painfully swollen knee will usually insist that he feels 'quite all right in himself', but when he is feverish or nauseated or when the pain is in his chest or abdomen, he will usually announce that he feels 'rotten' or 'seedy'. In New Guinea, patients suffering from serious generalised disease talk of themselves as 'ruined'.

Evans-Pritchard distinguished between Azande remedies

which were directed at the illness itself and those which were designed to counteract their human cause, but he regarded both groups as being essentially magical in character. There were certain exceptions — cleaning a wound with hot water, for instance, was common sense, and certain herbs probably had an antiseptic effect — but in most treatments there was no apparent relationship between what they did and what they hoped to bring about. The remedies seemed to be prompted by other considerations, and he called them magical to distinguish them from procedures which had an obviously practical justification.

This distinction applies to more than medical procedures. When a Zande craftsman makes a pot or a granary, most of what he does is obviously related to the final result. The techniques are self-explanatory. But the ritual precautions with which he sometimes accompanies the performance — spitting, whistling or muttered incantations — have a different significance altogether, although the craftsman himself regards them as an essential contribution to a favourable outcome. Such procedures merely accompany Zande craftsmanship and agriculture; but they constitute Zande medicine almost entirely. Admittedly, there are certain remedies, chiefly those used to treat mechanical disorders, in which the practical aspect overshadows the magical one. The treatment for sprains and broken bones, for instance, is nearly as technical as the procedures used for repairing broken granaries; in fact, there is a special clan called the Amozongu whose members specialise in these techniques and who are regarded by the tribe as straightforward craftsmen. But when it is a case of disease rather than injury, the magical character of the treatment prevails.

The definition of magic, medical or otherwise, has always proved very elusive. Unlike religion, it does not consistently address itself to any supreme authority, and the fact that it involves technical procedures distinguishes it from the unmediated supernaturalism of miracle-working, just as the fact that its procedures are not intelligibly related to a practical outcome distinguishes it from straightforward technology. There is no single feature by which magic can be recognised; there are, instead, a cluster of characteristics, some but not all of which have to be present before any particular procedures can properly be called magical.

One of them is the fact that the action is performed by someone who is acknowledged to be a magician. The community does not require a craftsman or a technician to show any peculiar characteristics over and above his proven ability to accomplish his chosen task, but the reputation of a magician depends not so much on what he brings about, as on the personal aura which he projects. He usually has distinctive features rather than technical qualifications, though formal apprenticeship may also be involved. Magicians are often odd in some way — they may be blemished, crippled, exceptionally old or strangers to the community. Certain occupations are traditionally credited with magical power: midwives and grave-diggers who usher people in and out of existence, and barbers who traffic in substances half-way between the living and the dead, like hair and nail-clippings, are likely candidates to be magicians.

But since a perfectly ordinary individual can practise magic, the social identity of the magician is only part of the story. The details of the action may be just as important. Unlike science, technology or craft, magic loses its credibility unless it is performed in certain places traditionally associated with the art. A blanket can be woven wherever it is convenient to set up a loom; pottery is usually performed where there is a plentiful supply of clay and water. But when magic is practised the locations are chosen with pedantic care: cemeteries, crossroads, the edges of woods, rubbish heaps, thresholds, rooftops and the boundaries of villages. Timing is just as important and equally unintelligible: the magician usually conducts his rites at midnight, dawn or dusk, whereas, unless a scientist is interested in observing an eclipse or measuring air temperature just before sunrise, he is indifferent to the hour at which he performs his experiment.

Sometimes the most distinctive feature of magic is not place or time but the way in which the magician prepares himself. A modern surgeon scrubs his hands and uses sterile gloves, because there is a clearly established relationship between dirty hands and septic wounds. The preparations for magic are much less understandable. A magician may abstain from certain foods or may fast; he may anoint himself, throw ashes in the air, burn candles or blow whistles. And while it is

Cooking and handicraft are often accompanied by magic but, unlike the magical procedures which are used to guarantee their success, the handicraft of the cook and potter is intelligibly related to the immediate outcome.

almost impossible to see how such preliminaries could influence what happens next, both the magician and his clients insist that the success of the enterprise depends on their correct observance, and failures are invariably attributed to their omission.

One of the most consistent features of a magical performance is the ritual use of words or slogans — something which is conspicuous by its absence from procedures which are obviously technical in character. Surgical operations are not usually performed to the accompaniment of rhymes or ditties: if a surgeon hums to himself or recites bawdy limericks under his breath, he does so to steady his nerves, and if he says anything else it is because he is trying to communicate with his staff. When the magician speaks, however, he is addressing neither himself nor his assistants: he is talking to, or rather, at, his materials. In fact, the language of magical spells is deliberately uncommunicative. The incantations are mumbled or muttered and, even when they can be overheard, they usually sound nonsensical — words may be spoken backwards or they may, as in the familiar 'abracadabra', just be salads of syllables. This is partly because magic is a secret art, and because some of its power depends on mystification. But it is also because language is being put to a totally different use: not to convey meaning, but to bring about effects. Spells are not communiqués: they are gestures. For the magician, saying or uttering is another way of doing. His speech is coercive and designed to bring things about. As far as he is concerned, what he says materially intervenes in the processes which he is trying to influence, and it is this aspect of the performance which makes it so unintelligible to someone unaccustomed to the use of magic. The civilised Westerner does not expect to make things happen by uttering words — not because he has tried and failed miserably, but because for him it would not even be worth the effort. He does not credit spoken words with material effectiveness. Even for the magician, the casting of spells is not the typical way of making things happen. When he resorts to his ordinary domestic tasks, he conspicuously abstains from such utterances, and if his cookery is occasionally accompanied by incantations, he takes care nevertheless to chop his vegetables and boil his meat in a conventional way: he would not expect the ingredients to render themselves into an edible meal simply by being addressed. The point is that, although it is impossible to give a reliable definition of 'making something happen', there are at least typical examples of doing so, and the uttering of spells does not count as one of them.

The same principle applies to the material procedures of magic — both to the substances which are chosen and to the way in which they are employed. In medical magic, the ingredients may not even be administered to the patient himself, but blown to the four winds, sprinkled round huts, or hung up on trees — from which one can only conclude that the medicine is not visualised in any obviously practical way. Even when the remedies are administered by more conventional routes — taken by mouth, rubbed on the skin or given by suppository — the substances used do not seem to be chosen with a view to their practical utility. It is not that they are inert or useless by definition: in fact, traditional remedies have sometimes formed the basis of scientific pharmacology — digitalis, quinine and opium all took their origin from folk medicine. But the original choice of such materials seems to be guided by principles which are altogether different from the ones which the believer in magic would use for his ordinary tasks. A tribal potter may try to guarantee the success of his next batch by sprinkling the clay with herbal juices, and it is difficult to see how such a baptism could influence the outcome. But in choosing clay rather than old clothes to make his crockery, the relationship between ends and means is immediately apparent.

The practical utility of magical remedies may be in doubt, but the ingredients are not chosen at random. In almost every case, it is possible to recognise some guiding principle. The English anthropologist Sir James Frazer identified at least two such principles: contagion and sympathy.

By 'contagion' he meant the magical power of physical contact. He noted that many folk remedies consisted of materials which had once been in close contact with the patient's body, or might even have been part of it. Sorcerers who were trying to make people ill and magicians who were trying to cure them often manipulated and recited spells over hair, nail-clippings, used bath-water and shreds of cast-off clothing. The magician does not, as Frazer supposed, confuse the part with the whole, but visualises and tries to exploit some invisible affinity which survives material separation. Such practices are not confined to primitive communities. In the seventeeth century the English scientist Kenelm Digby published a treatise on the so-called 'weapon salve', in which he claimed that it was possible to relieve an injured warrior by wiping his blood off the weapon which had inflicted the wound and dissolving the gory fabric in strong acid. Francis Bacon seriously discusses a cure for warts in which an apple is nailed in the sun after having been wiped upon the affected part. As the fruit withers, the wart is supposed to dwindle.

The belief in sympathy based on resemblance is even more widespread. Many of the ingredients or objects used in magic bear a superficial resemblance to the system or process on which they are supposed to work. Again, this is not, as Frazer assumed, because the magician in his infantile simplicity confuses things which resemble one another, but because he regards similarity as the outward sign of some underlying affinity or sympathy. A red herb may be used to counteract chronic haemorrhage because the colour which is shared by the plant and the blood implies some deeper connection

Magic often depends on the identification of resemblance or similarity. The problem is that there is no end to the number of resemblances which can be established. In this collection for instance you could group together all things that were blue – other things made of plastic – cylindrical things – toilet requisites – all things that were made of metal, etc, etc. Find some more.

between the two. When a Zande magician treats an epileptic patient by administering the ashes of a charred monkey skull, it is because he recognises some manipulable connection between the head of his patient and the skull of a monkey. For someone committed to magical thought, the universe is a network of analogies and sympathies.

The magician differs from the faith-healer and resembles the scientist in basing his practices on observation. He examines the world around him, recognises similarities and acknowledges distinctions. But that is as far as the connection between magic and science goes. For the magician, similarity is an unquestionable sign of underlying identity and automatically implies that the manipulation of one thing will inevitably cause changes in anything which resembles it. The difficulty is that there is always a respect in which something could be said to resemble anything else. Depending on which factor one chooses, the list of things which blood could be said to resemble is literally inexhaustible. If colour is taken as the relevant feature, rust, roses or redcurrants would have to be included; but since blood is sticky as well as red, honey and resin would have a legitimate claim — and, since blood flows, rain water and milk would also have to be acknowledged. In fact, if mere resemblance was regarded as significant, one would be led to the unmanageable conclusion that everything was connected to everything else, and even magic would be brought to a standstill by the *embarras de richesse*. For the magician, there are no rules or principles for discriminating between significant and insignificant similarities; there is no rationale for deciding between the indefinite number of distinctions that can be made between observed things. The world is a panorama of qualities, visible or tangible, and the only coherence which he can perceive is the infinite number of contrasts and resemblances which he can recognise.

This is perhaps one of the reasons why it is so difficult to recognise the diseases that are discussed in pre-scientific treatises on medicine. Their authors are not unobservant, but the criteria by which they observe are very different from ours, and are never explicitly stated. There is no reason to believe that the powers of sensory discrimination have improved over the years: in fact, to judge by the number of different pulses recognised and reported by Chinese and

Roman physicians, one could easily get the impression that clinical finesse has deteriorated. And the same goes for pharmaceutical ingenuity. Medical progress has been marked by a distinct reduction in the number of recommended drugs, and whereas it would be difficult to carry an eighteenth-century pharmacopoeia, the formulary used by a modern physician can easily be slipped into his top pocket.

The difference is not merely a question of simplification, but of the principles by which the simplification has been achieved. For the modern physician, disease is something more than the sum of all the regrettable changes that might be noticed. For him, what is worth noticing are those changes which are consistently associated with other recognisable alterations in bodily feeling or appearance. Confronted by a skin rash, for instance, and nothing else, a conscientious physician might be at his wit's end deciding what aspect of the rash he should notice and report. Should he mention the fact that the eruption has a scaly surface, or that the pimples are surrounded by a dusky halo: and how about the relative size of the spots? However, if he becomes aware of the fact that rashes of this general sort are also associated with painful swollen joints and that the scaliness of the rash is an invariable concomitant of the arthritis, then that particular aspect of the rash will gradually assume more importance. When a third symptom or sign is found as a frequent associate of the other two, the discrimination becomes even more refined.

The introduction of post-mortem examination carried the process of selective discrimination even further. With the discovery that internal organs could exhibit morbid changes, and that these were consistently associated with the clinical signs discovered in life, the picture of disease began to acquire a stereoscopic character, and this in turn made the clinical discriminations more stringent. For instance, before anyone thought of opening the dead body there was no limit to the number of ways in which abdominal pain could be described. But once autopsies were widely practised, it became apparent that only certain aspects of the pain were relevant to what was going on inside. Morbid anatomy therefore made an enormous contribution to the clinical description of illness, simplifying the number of possible distinctions and creating a manageable inventory of interesting bedside observations.

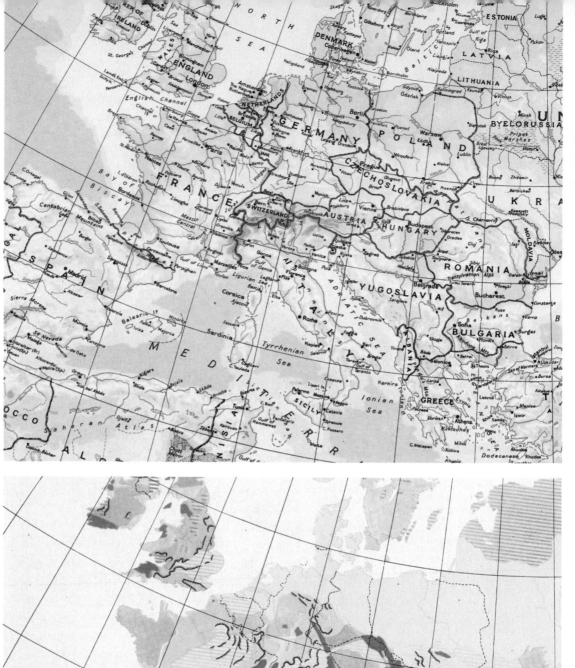

Giovanni Battista Morgagni.

Marie François Xavier Bichat.
(A. Coupé)

The man responsible for introducing these examinations was the eighteenth-century Italian physician Morgagni, who examined and cross-questioned his patients during life and then took special pains to record the appearance of their internal organs once they had died. In addition to refining the clinical description of disease, Morgagni's commitment to autopsy introduced a new idea as to what a disease was. For him, the changes in the internal organs were not merely alterations on a par with the ones observed in life, they were signs of the underlying process. Disease was seated in the ruined viscera.

Morgagni's pathology was expressed exclusively in terms of organs. As far as he was concerned, disease afflicted the visibly recognisable objects of the interior. Illness arose in the liver, in the heart or in the lungs. In 1800, however, the French physician Bichat identified a different order of pathological susceptibility. For him, tissues rather than organs were the site of disease, and the body was a composite of various material textures. He pointed out that the serous membranes which lined the inside of joints were identical to ones surrounding the heart and lungs, and that, although they were anatomically separated, their physico-chemical identity made them susceptible to the same pathological influences.

Morgagni and Bichat's views did not contradict but complemented each other. Like successive pages in a geographical atlas, Morgagni's picture was like a political map of Europe, divided up into organ states, while Bichat's was like a geological map, showing sheets of sandstone extending across the political boundaries. Fifty years later a third plate was added to the atlas, when pathologists such as Virchow insisted that the cell was the essential site of the disease process. Biochemistry and sub-cellular pathology have added a fourth level of possible description. One does not supplant another, however. All four contribute to the modern view of disease.

Such analyses make sense only in the light of normal physiology. To observe that some organ, tissue or cell has changed is useful only when one can say what function has thus been disturbed. By the end of the nineteenth century, pathologists such as Cohnheim were beginning to understand that pathology was nothing more than disordered physiology, and that many of the clinical features of illness represented

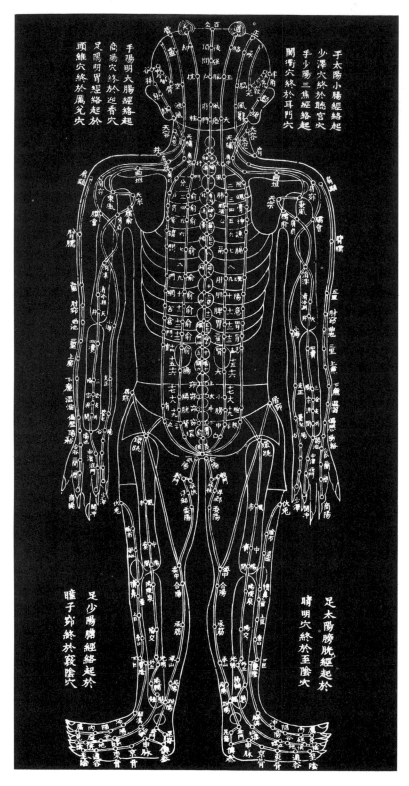

Maps do not necessarily represent an underlying reality. The acupuncturist justifies his practice in terms of a hypothetical system which has no independent evidence in favour of it. There is no end to the number of channels and tubes which can be drawn on to an outline of the human body but unless their existence can be systematically verified the diagram conveys no information. The same principle applies to the chart of moles on the human face and the supposed relationship between such blemishes and the system of astrology. (Chart of the face from R. Saunders, 'Physiognomie')

attempts on the part of the body to re-establish its balance and self-control. Admittedly, some of the most important advances in medicine have been the result of identifying and forestalling some of the agents responsible for the pathological changes which had been noted by Morgagni and Bichat — the recognition of the role of micro-organisms, for instance, and their control by asepsis and antibodies. But the most significant advance of all has been the recognition of the process by which the body helps itself and the tactful reproduction of the living system's own medical wisdom.

Greek monument to a
physician

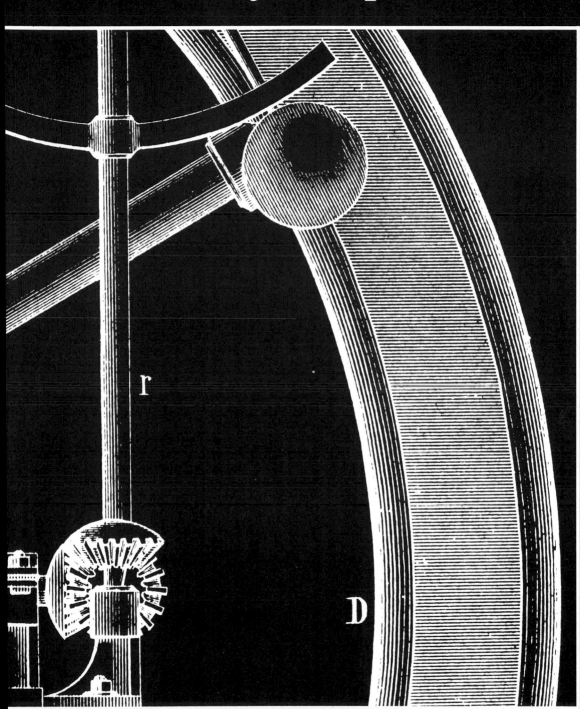

BY THE TIME ANYONE FEELS ILL ENOUGH TO CALL IN A DOCTOR, he has already been receiving free treatment from a private physician whose personal services have been available to him from the moment of his conception. By inheriting the premises in which we are condemned to spend the rest of our lives, we are born into a hospital whose twenty-four-hour services are, paradoxically, designed to overcome and counteract the risks of living in such a dangerous tenement. It is a hospital staffed by its only patient, and although we take no conscious part in our own therapeutic activities, the fact that we have ourselves on call around the clock means that we can overcome most common emergencies without having to summon outside help.

Prevention is, of course, better than cure, and ideally one would live in a world so free of risk that prevention and cure would be equally unnecessary. But it is hard to imagine what such an environment would be like. The womb is the nearest we ever come to it and, like the womb, such a world would be so monotonous and so unchallenging that any species which grew accustomed to it would soon lose the ability to survive anywhere else. The most versatile and ambitious species are those which have evolved mechanisms capable of recognising and facing threats before they have had a chance to inflict expensive and possibly irreparable damage.

Most of the protective mechanisms I am going to discuss are reactions or responses of one sort or another: they are called into existence when a threat makes its appearance. Once they have succeeded in forestalling the threat, they automatically subside — ready to issue forth again whenever the danger recurs. Creatures which have a large repertoire of such mechanisms are much more favourably placed than animals which rely on permanent protection. Shellfish, for example, depend for their safety, such as it is, on heavy mechanical armour, which is cumbersome and only occasionally useful. When an animal is as small as a water flea, the weight of its shell does not limit its mobility, but when it is as large as a turtle, most of its muscular energy is spent on moving its safety from one place to the next. Anyway, no armour is foolproof, and animals which rely on heavy shields are no match for ones which have invested in nimbleness, intelligence and ingenuity.

The same principle applies to unbreakable bones.

On pages 104–5
*Watt's centrifugal governor
(see also pages 138–9).*

The nervous system may become so adept and versatile that its owner may take a positive pleasure in pitting it against dangerous risks.

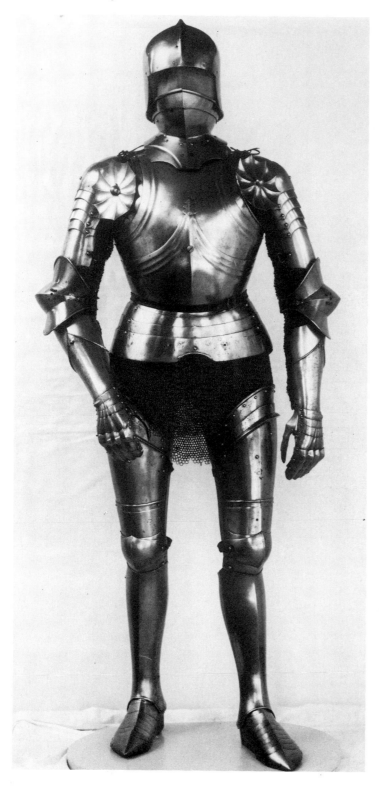

Heavy shells and unbreakable
bones are expensive and
unprofitable investments. A
nimble body and a versatile
nervous system pay better
dividends.

(Left   Suit of armour,
The Wallace Collection)

When an organism is as small
as this its shell affords some
security without limiting
mobility.

Theoretically it would be possible to construct a skeleton which could resist all but the most extreme mechanical stresses. But the weight of such an apparatus would be quite prohibitive, and in the long run it is more profitable to have a nervous system which can anticipate dangerous falls, complemented by bones which can rapidly repair themselves in the event of unavoidable accident. As a general principle, living organisms distribute their insurance funds between a nervous system capable of forecasting threats, and tissues whose active and continuous growth enables them to repair unpreventable damage.

The way in which an animal reacts to a threat depends to a large extent on how far away the threat is when it first makes itself known. Remote dangers call forth a totally different type of behaviour from ones which first appear on or near the surface of the animal. The farther away the threat is, or the sooner it is recognised, the more opportunity there is to analyse the situation as a whole. When a predator announces

Discretion may be the better part of valour. One way of avoiding injury is to inherit the appearance of one's own background.

Opposite and above
*A nervous system which can
recognise and react to threats as
and when they occur.*

its distant presence, the animal has time to rearrange its whole
attitude and plan the most profitable course of action. If
the opponent is larger, fiercer or faster, flight may be the
best form of protection. If the opponents are more evenly
matched, particularly if they are members of the same species, it may be more appropriate to put up a frightening
show of aggressive behaviour — or a display of submission
may be just as effective, since it often calls off the fight; but
if battle is joined, the weaker contestant immediately arranges
itself to minimise the expected injury: it folds up and offers
the least vulnerable surface. These reactions are reinforced
by physiological changes in the bloodstream. Alarm causes
the automatic secretion of the hormone adrenalin, which,
when released into the bloodstream, speeds the heart, raises
the bristles, increases the blood sugar, and generally puts
the animal in a favourable condition for either fight or flight.

As the threat approaches, the reactions become brisker,
simpler and more localised: instead of involving the animal's
whole attitude, they are limited to the threatened part itself:
eyes blink and limbs flinch. The animal is, in other words,
protected by concentric rings of vigilance. Vision, hearing and
smell anticipate distant threats in time to plan strategic action.
The behaviour connected with these sense organs is therefore
judicious, hesitant and versatile. But anything which succeeds
in sneaking past or breaking through the early-warning system poses a much more urgent problem. With the immediate
threat of injury, there is no time to consider alternative actions, and the response must be simple, unhesitating and
automatic.

The sequence repeats evolutionary history in reverse: urgency reduces behaviour to a simplicity which is comparable
with that of our primitive ancestors. This principle also applies to the pursuit of satisfaction. The reactions which fulfil
and consummate satisfaction are much less versatile and
much more localised than the ones which are used to obtain it:
swallowing food is a much more automatic affair than hunting
for it; orgasm is much more stereotyped than courtship —
which is probably why pornography is so much less interesting than literature.

There are some threats which, because they are too small to
be seen or too quiet to be heard, always arrive at the surface

Opposite
Protective flinching is so deeply engrained in the human nervous system that it is still performed in the face of overwhelming odds.
(Above *The shooting of Harvey Lee Oswald, President John F. Kennedy's assassin, photograph © 1963 by Bob Jackson, Dallas Times Herald.* Below *The Hungarian uprising, 1956. Police being shot. Photograph by John Sadovy*)

unannounced and menace us before we have had time to prepare ourselves. Such dangers can usually be handled by short-range tactical mechanisms. The scratch reflex, for example, is so automatic and so primitive that animals can display it even when they have lost higher parts of the nervous system — a frog which has had its brain destroyed and is completely insensitive to visual or acoustic stimuli will nevertheless reach up and try to wipe away a cutaneous irritant. The eye has an even more efficient arrangement. Its surface is so delicate and the transparency of the window so important that it anticipates injury with a non-stop laundry of tears and blinks; when an irritant actually lands on it, the output of tears increases, and the eyelids go into protective spasm. These mechanisms are so automatic that it is easy to take them for granted, but if illness dries up the tears and paralyses the blinking, the transparent surface soon ulcerates.

The lungs and intestines are also guarded by reflex mechanisms. Although these passages are folded away inside the torso, the membranes which line them are just as much part of the outside surface as the skin is. Fortunately, they have comparatively narrow entrances, and the disproportionate sensitivity of the nose, tongue and lips makes it easy to supervise their safety. Although the intake of food endangers the uptake of breath — to reach the stomach, food has to pass over the top of the windpipe — confusion is avoided by the subtle co-ordination of the two functions. The act of swallowing automatically postpones breathing and at the same time raises the larynx, so that the entrance to the lungs is brought under a little flap or lid just behind the tongue called the epiglottis — you can see this action quite clearly if you watch someone's Adam's apple when he swallows. If this co-ordination is paralysed, as it may be in cases of polio, for example, the patient is in danger of choking whenever he swallows.

This and other dangers call for a second line of defence. Any irritant which succeeds in entering the windpipe immediately excites the familiar cough reflex. The patient takes a breath, but as he starts to blow out, the larynx narrows, so that the act of expiration works for a moment against a strong resistance. When the air pressure inside the lung has been raised to a sufficiently high level, the larynx suddenly relaxes, and the explosive release of air blows the irritant out of the mouth or nose. This laryngeal action is extremely important: patients who have lost control over it may be able to puff and blow, but they can't raise the explosive power necessary to expel the foreign body.

A cough may be an irritating inconvenience, but it would be disastrous if the reflex failed. Doctors are sparing in their use of cough mixtures because, if it were extinguished, the unsuspecting lungs would lie open to the invasion of septic material. This is also why unconsciousness poses such a serious threat to the lungs. Loss of the cough reflex is now considered one of the most serious hazards of head injury: the victim of a road accident runs as much risk from drowning in his own secretions as he does from the immediate effects of the injury itself.

Recognition of this simple fact has led to a dramatic fall in the mortality of unconscious patients. It is amazing how easy it is to save life by turning the unconscious person on his front. This prevents his tongue from falling back into his throat and allows blood, mucus and saliva to run out of his lungs and into his mouth. If the unconsciousness persists, it may be advisable to perform the operation known as a tracheotomy, which isolates the windpipe from the mouth and the throat and at the same time makes it more easily accessible to the nursing staff,

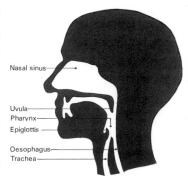

Nasal sinus

Uvula
Pharynx
Epiglottis

Oesophagus
Trachea

Above   *Breathing*
Below   *Swallowing*

A post-operative tracheotomy.

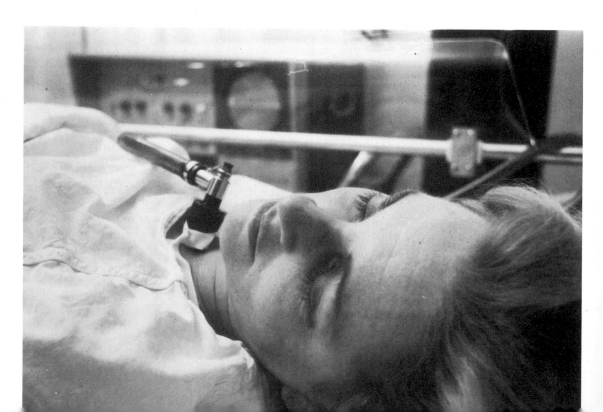

Control room, NASA.

who can suck out secretions before they have had time to accumulate. In other words, it enables the doctor and his assistants to stand in for the cough reflex. By understanding the purpose, we can reproduce the function.

The management of the unconscious patient is one of the unexpected benefits of successful anaesthesia. Anaesthetics were introduced in the middle of the nineteenth century to make the patient unaware of surgical pain, but they also gave doctors an admirable opportunity to observe the lethal negligence which is inevitably associated with loss of consciousness. The anaesthetised patient has been deliberately stripped of his protective vigilance and reduced to the status of a physiological robot, wide-open to all the threats the world has to offer. The cough reflex is abolished, and, unless the windpipe is isolated, the patient is in danger of asphyxiating during the operation — especially if he vomits. Once doctors recognised that unconsciousness as such represented a serious threat to survival, it became apparent that what had been learnt from the management of anaesthetised patients could be applied to those suffering from stroke and serious head injuries. Today, the unconscious patient is treated as if he were an astronaut in orbit: his safety is monitored and managed for him; he is carefully steered through all the natural risks until he is ready to link up again with his own command module and assume these responsibilities for himself.

The cough reflex is only one element in an elaborate system of respiratory protection. It is the most noticeable because it is such a strenuous action and because the cougher is usually aware of the stimulus which provokes it. But the surface of the respiratory tract is protected against quieter and more chronic threats as well. It is lined with cells, many of which are capable of pouring out a protective layer of mucus which smothers irritants and bacterial invaders before they have a chance to penetrate the living membranes. Other cells are equipped with mobile bristles whose concerted action wafts the mucus up to the mouth, where it can be swallowed or spat out. Chronic irritation may paralyse this microscopic escalator, which is one of the reasons why heavy smokers are so susceptible to repeated bronchial infection.

The risks associated with respiration are more immediately threatening than those associated with eating and drinking: the need for an uninterrupted supply of oxygen is so vital that anything which obstructs it represents an urgent threat to life. Nevertheless, there *are* risks in taking nourishment, and we protect ourselves against these as far as possible. One again, the versatility of the precautions depends on the proximity of the threat: we sniff and look at our food, and with the help of intelligence and training we learn to reject anything which seems suspect; if something escapes the vigilance of our eyes and nose, our sense of taste may still save us from swallowing it. None of these devices is foolproof, however, and swallowing is an irreversible act. Therefore, the gastro-intestinal tract is equipped with a counterpart of the cough reflex.

Anyone who has had a serious attack of vomiting would probably find it hard to appreciate its usefulness. But it is just as indispensable as coughing and just as complicated: it is not simply intestinal action in reverse. In 1813 the French physiologist François Magendie published a pamphlet rather unpromisingly titled *Mémoire sur le vomissement*, in which he described an experiment which proved that vomiting is not performed by the stomach as such, but by the voluntary muscles of the diaphragm and the abdominal wall. He administered an intravenous emetic to two experimental animals. In one of them, he had paralysed the muscular diaphragm,

The waves that move across a field of corn resemble the movement of cilia. (Van Gogh, 'Wheatfield with Lark', National Museum Vincent van Gogh, Amsterdam)

Mobile bristles or cilia on the respiratory surface protect the lung from foreign silt.

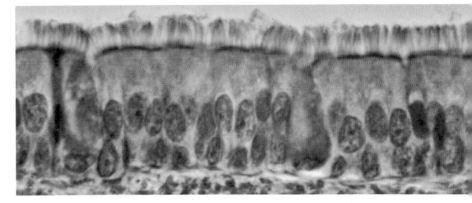

The bristles' internal structure can be seen under the electron microscope.

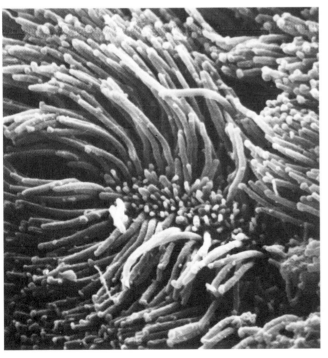

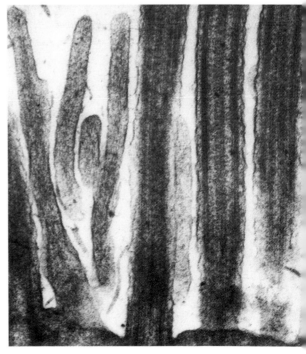

leaving the normal stomach in place. In the other, he left the diaphragm intact but replaced the stomach with a thin-walled artificial bag. The animal with the intact diaphragm was still capable of throwing up the contents of its artificial unmuscular stomach, but no amount of emetic produced vomiting in the animal with the paralysed diaphragm. The conclusion was inescapable: the muscular stomach plays no projectile part in the act of vomiting. The emetic substance excites the vomiting centre in the brain, and as a result of this the diaphragm contracts and plunges into the abdominal cavity, rather like a piston head moving into a cylinder. Simultaneously, the muscles of the abdominal wall become spasmodically rigid. These two actions together raise the pressure inside the abdominal cavity, and since the stomach has an open chimney, its contents are automatically projected.

Vomiting, like coughing, conscripts and redirects the muscles normally engaged in breathing in order to overcome a more pressing emergency. When it has fulfilled its protective task, the reflex disestablishes itself and the muscles are reassigned to their regular function. Such versatility, which presupposes a nervous system endowed with an unfailing sense of biological priorities, is characteristic of the economy with which the body defends itself. Instead of depending on a large number of separate mechanisms, each one of which is exclusively reserved for its own particular type of emergency, the body improvises responses to the threat of injury by assembling new combinations of pre-existing functions. The manoeuvres by which an individual corrects a dangerous stumble shade imperceptibly into the efforts required to maintain the upright posture. The English neurologist Charles Sherrington pointed out that each step is a stumble caught in time.

The same principle of thrifty versatility applies to the elaborate adjustments the body makes as a result of injuries which it has been unable to prevent. Perhaps the most dangerous consequence of injury is the loss of blood. The efficiency of the heart and of the circulation depend on the maintenance of an adequate blood pressure, or fluid volume, and, if more than a pint or two is lost in a short space of time, the system goes into a state of almost irreversible hydraulic shock. It is like a rubber syringe which, if it is filled and then squeezed,

The French physiologist, François Magendie, 1783–1855. Preceded by François Xavier Bichat and succeeded by Claude Bernard, Magendie laid the foundations of modern experimental physiology. Although he recognised that life was more than the sum of its material parts he refused to explain vitality as the expression of an immaterial force. For him, even the human intellect had to be considered as the result of the action of the brain and was thus indistinguishable from any other phenomenon which depended upon organic action.

MÉMOIRE
SUR
LE VOMISSEMENT,

LU A LA PREMIÈRE CLASSE DE L'INSTITUT
DE FRANCE,

Par M. MAGENDIE,

Docteur - Médecin de la Faculté de Paris ,
Prosecteur à la même Faculté, Professeur
d'Anatomie, de Physiologie, etc.

Suivi du Rapport fait à la Classe par MM. Cuvier ,
Humboldt, Pinel et Percy.

A PARIS,

Chez CROCHARD, Libraire, rue de l'École de
Médecine, nº 3.

1813.

The title page of Magendie's 'Treatise on Vomiting'. Like Harvey's work on the circulation of the blood, Magendie's work exemplifies the so-called Hypothetico-deductive method. A conjecture is put forward, its implications are then put to the test of concise experiment.

produces a powerful jet of fluid; but if the bulb is not replenished the jet becomes feebler at each succeeding squeeze, and eventually the pressure falls below the level where it can overcome the resistance of the tubes through which it has to pass. In the body, if this happens faster than the system can replace the loss, certain vital organs begin to fail for want of oxygen and nourishment. The most delicate organ, the brain, fails first — which is why people feel faint and lose consciousness when they lose blood rapidly. Further loss endangers the kidney, which needs a high head of pressure to perform its essential function. Therefore, when all strategic and tactical measures have failed — when the skin breaks in spite of prudence, cringing, wincing and flinching — it is essential to have standby mechanisms which stop the flow of blood at the earliest possible opportunity.

One of these is the reflex shrinkage of the cut vessels themselves. Arteries, in particular, which have a thick muscular wall, can automatically shut down and limit the outflow. This is reinforced by an elaborate response on the part of the blood itself. Through a complex chemical mechanism, certain proteins in the blood are automatically transformed when they pass over a damaged surface. At the site of the injury, a skein of elastic fibres is woven in the blood. This rapidly forms a clot, which organises itself into a plug, preventing further bleeding. Although there are limits to the effectiveness of the plug, it deals adequately with most of the minor causes of blood loss. People whose clotting mechanism is impaired run the risk of harmful if not lethal bleeds unless they avoid the normal hazards of life. The hereditary disease known as haemophilia is the most famous example of such a condition, but there are many chemical factors involved in successful clotting, and haemophilia is only one of a number of 'bleeding diseases'.

Blood requires just as complicated a mechanism to prevent it from clotting as it does to make clotting occur. If it clots spontaneously, which is what happens in patients who lie sluggishly in bed after an operation, and is also one of the recognised risks of the contraceptive pill, the person runs the risk of plugging his or her own circulation from the inside. The clotting mechanism has to be in a state of hair-spring readiness so that it can provide a clot at the moment of injury

Opposite   *Each step is a stumble caught in time. (Marcel Duchamp, 'Nude descending a staircase', © 1978 A.D.A.G.P., Paris)*

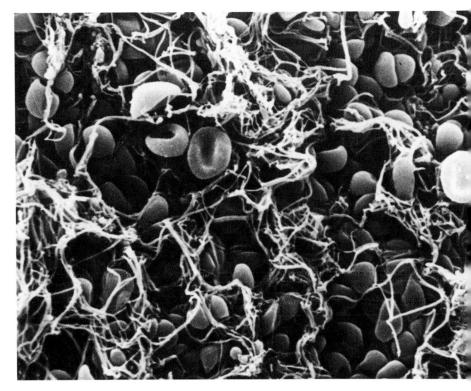

Red blood cells caught in a tangled skein of fibrin. In small vessels this mesh forms an effective block and prevents dangerous blood loss.

and no later, but it must also have an elaborate fail-safe device which prevents it from being launched inappropriately. It would be tedious to list all the factors which co-operate in this, but there are at least twelve stages which have to be gone through before clotting will occur and, if any of these fails, the mechanism aborts: it is rather like the series of orders, locked safes and sealed levers which prevent the accidental launching of a nuclear missile.

Although normal clotting is punctual and self-limiting, it prevents bleeding only in vessels where the pressure is comparatively low: in a major artery the clot is dislodged as fast as it forms and, unless a tourniquet is applied, the patient runs the risk of bleeding to death. The body is unable to stop a major haemorrhage, but up to the last moment it strenuously adjusts itself to the continuing loss by a process of reallocation. As the blood pressure starts to fall, organs which have a less urgent need for oxygen and nourishment are put on short rations: the arteries which let blood into the skin, for instance, shut down and limit the blood flow, shunting it to the more vital parts of the circulation. This is why patients with haemorrhage go white and feel cold to the touch.

In the early stages of haemorrhage these and other compensations are quite successful, but if the loss continues, or if it goes faster than it can be artificially replenished, it eventually overwhelms even the most stringent acts of redistribution, and the patient goes into the state known as shock. This should not be confused with the anxiety and mental distress which go by the same name. Surgical shock is what happens when the process of compensation fails to keep up with the rate of blood loss — when the heart can no longer maintain the pressure needed to irrigate the vital organs, and the circulation begins to collapse. The patient is deathly cold to the touch, his pulse becomes rapid and thready, and his blood pressure drops to the point where it becomes unrecordable. The circulation is now powerless to maintain itself by differential shrinkage, and unless the blood loss is replaced artificially by a transfusion the shock becomes irreversible: either the patient dies or his kidneys and brain suffer irreparable damage.

Shock is the consequence of an underfilled circulation. In the first instance at least, the system is failing for want of

In the effort to maintain its volume blood dilutes itself after haemorrhage by withdrawing fluid from the tissue spaces.

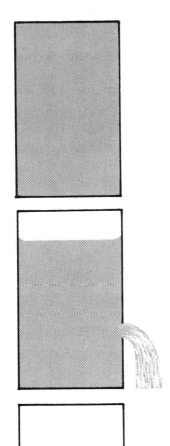

ISSUE FLUID

CLOT

volume, and the most important priority is the restoration of fluid bulk. If the loss is sufficiently slow, the body can do this on its own behalf by diluting the blood. Fluids are withdrawn from the tissues; there is an automatic reduction in the output of urine and an increased intake of water by the mouth — haemorrhagic patients are characteristically thirsty.

If the patient survives the period of shock, the volume of circulating blood returns to normal within twenty-four hours of a serious haemorrhage. But the rate at which new cells can be manufactured is much slower than the speed at which fluid can be restored, so the blood count is much lower than normal. Instead of having 5 million cells per cubic millimetre, the patient may have only 3 million: his blood pressure may be adequate, but he is now anaemic. The process doesn't stop there, however: anaemia acts as a stimulus to the tissues which produce new red cells. The marrow of such a patient begins to pour out more than the usual number. This process is elegantly self-limiting: the more successful it is, the less need there is for subsequent production, and when the red cell count has been restored to its normal level, the activity of the marrow subsides.

It does not, however, subside to zero. The only reason why the marrow is able to make up for unexpected losses of circulating red cells is because it is busily engaged in replacing the chronic losses of the normal circulation. The reaction to haemorrhage, therefore, is like the light touch on the accelerator of a rapidly moving car. Similarly, the way in which the blood vessels react to haemorrhage is indistinguishable from the variations by which they compensate for the normal stresses of daily existence. The reallocations which take place after a haemorrhage are simply dramatic versions of the ones which happen each time we rise from a recumbent position, when the body automatically adjusts itself to the force of gravity by constricting the blood vessels in the legs, thereby preventing too much blood from accumulating in the lower part of the body at the expense of the brain. This delicate mechanism becomes unpractised when the subject is removed from the gravitational field: when astronauts return to Earth after undergoing long periods of weightlessness, their cardiovascular nimbleness is so seriously impaired that they faint every time they stand up.

The most substantial advances in medicine are those that recognise and successfully reproduce the body's natural processes of compensation and repair. The successful treatment of haemorrhage is a vivid case in point. The most immediate threat to life is not shortage of blood, as such, but an inadequate volume of fluid. In the first stages of haemorrhage, when the circulation falters for want of sheer bulk, the doctor's first task is to reprime the pump before its mechanics fail irreversibly. The administration of plasma — the straw-coloured fluid in which the red cells float — or plasma substitute, is an acceptable stop-gap in the early stages, since it imitates the natural tendency of the body to restore the bulk of the blood at the expense of diluting it. But the transfusion of whole blood is obviously the treatment of choice, since it telescopes the two phases of natural recovery, restoring both the quantity and the quality of what has been lost.

The way in which the blood restores itself after haemorrhage is a specific example of the healing process by which tissues throughout the body repair injuries which the nervous system has been unable to forecast and prevent. Since blood is

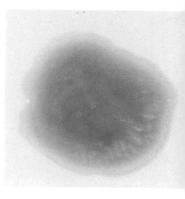

Laboratory test of compatibility. When red cells from the potential donor are suspended in the serum of the hopeful recipient they should remain evenly distributed. If the blood groups are incompatible the cells clump and disintegrate, producing dangerous reactions. All transfusions are preceded by grouping and cross-matching.

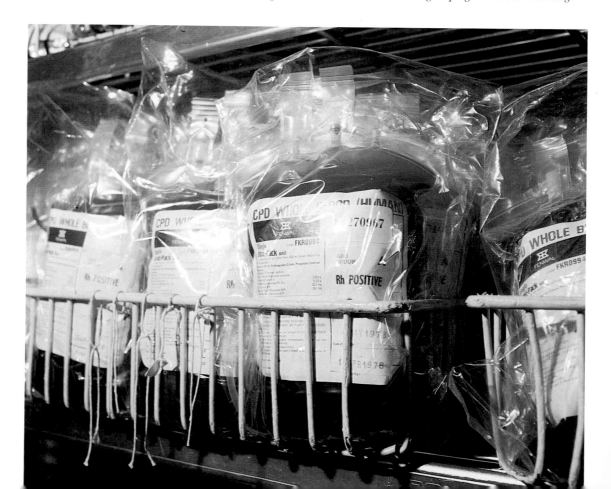

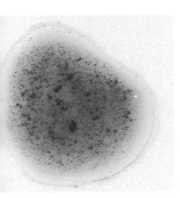

Right

The exchange of blood turned out to be much more difficult than this convivial fantasy suggests. Blood clots as soon as it is shed and the tubes block. In any case the acceptability of blood depends on the genetic compatibility of donor and recipient. Blood groups were not discovered until 1900.

Opposite

A living individual may not be able to hoard blood within his own body but a civilised community can do so on his behalf. A healthy person can spare at least a pint of what he has and although he may never receive payment either in cash or kind, he is amply rewarded by the reassuring knowledge that the contribution made by others establishes a public fund of blood upon which anyone can draw in time of need. A free transfusion service creates a 'gift relationship' between the members of a community and although donors and recipients are unknown to each other the fund which links them strengthens the vigour of the body politic.

on the move, it is difficult to think of it as a tissue and therefore hard to visualise the process of repopulation as an example of repair. However, the ability to heal and repair injury is shared by most of the body's tissues, although some are much more efficient than others. The reaction of the bone marrow is identical to that of any other tissue which is endowed with the capacity to restore the *status quo*. Ironically, the organ responsible for anticipating and forestalling damage is the least able to withstand it when it happens. While the skin, the gut, the bones are able to regenerate themselves after quite extensive injury, the central nervous system is unable to do so. It can provide scar tissue — that is to say, cells and fibres which restore mechanical continuity — but it cannot reproduce the tissue's original function.

Healing and repair are invariably preceded by a physiological overture known as inflammation. Since antiquity men have recognised that redness, heat and local swelling are associated with the occurrence of injury and infection, but until the eighteenth century these were regarded as painful ordeals and nothing more. It was the Scottish surgeon John Hunter who first suspected that inflammation, far from being an illness, was the first stage in recovery and repair, and with the invention of refined microscopes in the nineteenth century, scientists began to appreciate the subtlety of the process.

At the site of the injury, the damaged cells release powerful substances which cause a rapid local alteration in the behaviour of the blood vessels. At first, the vessels expand, allowing more blood into the affected part. This accounts for the heat and the redness. The same substances also change the permeability of the local blood vessels, so that serum, the part of the plasma which is left after the clot has been precipitated, begins to ooze into the surrounding spaces. This leakage is simply an exaggeration of a process which is going on all the time: but under normal circumstances the fluid is reabsorbed by the lymph vessels as fast as it is exuded. When injury occurs the serum leaks faster than it can be reabsorbed, producing local swelling.

This sequence of events is repeated wherever injury is inflicted, inside the body or on the surface. On the surface of your own skin you can actually observe the large-scale changes. They were first described by the English physician Thomas Lewis during the middle years of this century. In a great work entitled *The Blood Vessels of the Human Skin and their Responses*, Lewis showed that a consistent sequence of events followed even the mildest injury to the skin. If you draw a firm line down the skin of your forearm with the edge of a ruler, the first thing you notice is a white line as the local blood vessels shrink in anticipation of blood loss. After a few moments, a bright red streak appears: the blood vessels are expanding in order to allow freer circulation through the area. A moment later, you can see a distinct ridge along the red streak where the vessels have leaked fluid into the tissue

Lewis's inflammatory response could sometimes be evoked by trivial stimulation. It is possible to raise weals on the skin of such patients by stroking them lightly with the tip of the finger — so-called dermatographia.

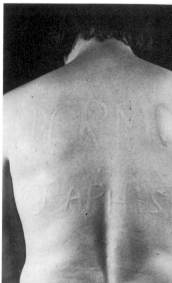

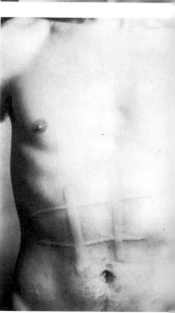

spaces. Lewis believed that the reddening and the swelling were due to the release of something he called 'H substance', and he proved this by a very simple experiment. He drained his arm by holding it upwards for a few minutes and then applied a tourniquet to prevent any more blood from reaching it. He found that it was no longer possible to obtain the raised red weal, but that this appeared as soon as the blood was allowed back into the arm on releasing the tourniquet. Lewis concluded that the inflammatory 'H substance' must have been released by the injury, but that it was unable to show its effects until the blood had flowed back into the arm. Since the red weal appeared immediately after the tourniquet was released, the 'H substance' must have prepared the blood vessels already, and only an inrush of blood was needed to show up the effect. It was as if he had drawn a streak on a photographic plate but had been unable to show it until it had been flooded with developer. We now know that a significant fraction of Lewis's 'H substance' is made up of something which was later identified as histamine. This is a simple organic material which is stored in specialised cells found in every tissue of the body. When it is released by injury — mechanical, thermal, bacterial or chemical — it invariably produces the same response in the small blood vessels with which it comes in contact: it causes them to dilate so that more blood flows through the affected area, and it increases the permeability of their walls, so that the serum can ooze into the surrounding tissues. By doing this, the histamine sets the stage for the initial phase of healing and repair.

First, the debris of irreparably damaged cells has to be removed. We have retained in our bloodstream active representatives of our original one-celled ancestors, white blood cells, and these stroll freely through the circulation, ready to make themselves available when the inflammatory call is raised. This sends them worming their way through the walls of the expanded vessels, attracted by the chemical lure of the substances released by the injury, and they remove the debris and bacteria by engulfing them. These cells sometimes accumulate in such numbers that they add to the swelling and are eventually discharged in the form of pus.

The pain and stiffness which usually accompany inflammation are not just incidental features of the process.

When the damaged tissues are still in the early stages of repair they are very sensitive to mechanical interference. In the early 1860s, an English surgeon called John Hilton recognised that the pain encouraged the part to immobilise itself and thereby protect the regenerating structures from unnecessary movement. He pointed out, further, that the stiffness of an inflamed joint splints it while it is being repaired; in peritonitis the painful rigidity of the abdominal wall helps to localise the infection and protect the underlying processes of inflammation. By enforcing rest, pain assists recovery.

Under normal circumstances, inflammation is set in motion by some injury or infection and the intensity of the process is proportional to the severity of the assault. In other words, the amount of histamine which is released is proportional to the damage inflicted. Some people, however, are hypersensitive. Their tissues tend to release large amounts of histamine as a result of trivial stimuli. The mildest example of this is the condition known as dermatographia, in which the lightest stroke on the skin will reproduce Lewis's flare and weal. Such people can write their names on their skin with the tips of their index fingers. The symptoms of such abnormal sensitivity may be much more severe and inconvenient than this, however: hives and asthma are two of the more familiar examples. As one might expect, some of these conditions can be treated by the administration of anti-histamines, which block the release of the inflammatory substance.

Usually inflammation is self-limiting. Like the other processes I have discussed, it is inhibited by its own success, and as it accomplishes its appointed task it gradually subsides so that the next stage of healing can begin. But sometimes the process becomes drawn-out and a state of chronic inflammation ensues. The irritant or infection proves intractable, so that the inflammation establishes and re-establishes itself without fulfilling its purpose, in which case it becomes an illness in its own right, and the lesions which result assume a consistent appearance. This is what happens in, for example, tuberculosis.

Sometimes the inflammation perpetuates itself because the body has lost the mechanism which switches it off and, once the vicious circle is established, the person suffers all the ordeals of pain and swelling without the consequent benefits

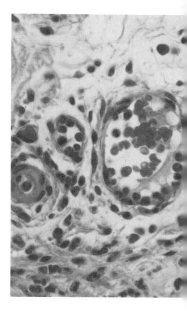

In the first stage of inflammation the capillary walls become sticky and the white cells become attached prior to worming their way into the surrounding tissues. The red corpuscles can be seen at the centre of the vessel.

The white cells engulf foreign material.

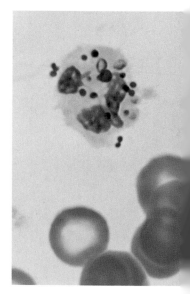

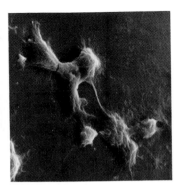

At a later stage of inflammation special cells called macrophages enter the scene. They engulf debris and then transform themselves into the tissue which makes up a scar.

The tubicle bacillus sets up chronic inflammation and instead of setting the stage for repair and recovery the inflammation process becomes a disease in its own right.

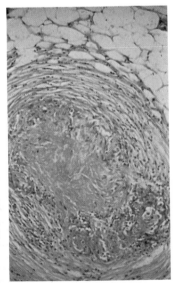

of healing and repair. Nevertheless, in spite of these risks, it would be disastrous if the process of inflammation were to be prevented or paralysed. In fact, drugs like cortisone, which have a specific restraining effect on inflammation and are often usefully employed for that purpose, may expose the patient to the risks of uncontrollable infection: a patient may be relieved of the painful inflammation of arthritis but made vulnerable to infectious diseases.

This is also what happens when a doctor uses cortisone to suppress the inflammatory response to a much-needed skin or organ graft. In the course of its development, the foetus compiles an inventory or identity card of all its own materials. When this inventory is complete, all other biological material is regarded as non-self, and whenever such an alien substance is introduced, an elaborate process of inflammation and antibody production ensues, and the intruder is eliminated. This arrangement, designed to protect us against viruses and bacteria, is so indiscriminately efficient that it also attacks organs and tissues which have been introduced for our good. To try to overcome this prejudice against foreign immigrants, the physician uses drugs or radiation to paralyse the mechanism which recognises and then rejects foreign material. When a patient is dosed with cortisone or X-rays, the basis of natural immunity is suspended and the indispensable foreigner is slipped past the immigration authority. Unfortunately, the inattentiveness of the tissues allows all sorts of undesirable aliens to enter as well, and patients whose bodies have been inveigled into accepting a graft are equally susceptible to invasion by less helpful material. Transplant patients have to be nursed in scrupulously sterile wards and prevented from meeting their septic loved ones. If you remove the natural processes of protection, you must restore the patient to the womb-like monotony which exists only in special hospitals and on the far side of paradise.

Once inflammation has completed the task of clearing the debris, the second phase of healing and repair can begin. In some tissues, this is accomplished by extending the processes by which they maintain themselves from moment to moment. Organs or tissues which encounter an unremitting wear and tear are already renewing themselves at a rapid rate, and since an orderly replacement of cells is already in hand the extra effort needed to replace unexpected or abnormal losses is almost negligible by comparison. Such tissues retain their form by ceaselessly renewing their substance, and the process of healing shades imperceptibly into this background of chronic maintenance.

The problem, then, is not how healing is set in motion, but how the continuous process of growth on which it depends is prevented from running away with itself. The mechanisms which serve this self-control are only just beginning to be recognised, but it appears probable that growing tissues are endowed with a self-regulating device which automatically keeps the growth at a level which maintains the tissues in the required form, and that, once foetal development is complete, cell-division metronomically ticks over at the rate needed to maintain the adult size and structure. Built into this mechanism, it seems, is the ability to recognise gaps or losses due to injury. The normal growth-rate is accelerated in proportion to the loss, and then switched or tapered off as the deficit is made good.

Not all tissues turn themselves over at such a rapid rate, but most of them retain the capacity to renew the process by which they came into existence whenever mechanical injury inflicts a detectable flaw in their continuity. Liver and kidney, for example, do not continue to grow and renew themselves in the way that skin or bone marrow do. Presumably there is no need for them to do so, since they are not exposed to such unremitting loss. Nevertheless, they can restore their own architecture whenever injury is inflicted.

The success of such an insurance policy depends upon an accurate and subtle restraint of the cellular process which it exploits. The fact that cells are poised to multiply themselves in anticipation of injury, and that this potential is held in check by constitutional restraints which are built into the genetic instructions of each cell, means that there is always the risk of anarchic proliferation. When the self-regulating mechanism is impaired and the growth-rate begins to outstrip what is demanded by daily erosion or by acute injury, a condition of 'new growth' sets in, and cancer ensues.

Healing, repair and maintenance are part of a continuous spectrum which extends to include the process by which the fertilised egg achieves the organised structure of the adult. All

If you remove or diminish the natural protective systems of the body a patient may be forced to live in a risk-free environment created by doctors.

of them exhibit and exemplify the fundamental principle known as homeostasis, through which a genetically determined norm is first achieved, then maintained, and finally restored in the face of unexpected injury.

Although the principle of homeostasis is most vividly exemplified in the case of physical injury, it applies to anything which has a bearing on the welfare and efficiency of living systems. In fact, biological matter is so delicately poised that almost any change in the environment represents a challenge to its survival. The metabolic processes of a living cell are highly susceptible to changes in temperature, for example.

131

Each type of organism has an optimum temperature at which its metabolism works most efficiently; if it rises to a certain point above that, the chemical transactions speed up so much that the system runs the risk of consuming itself; if it falls too low, the chemical processes slow down and eventually stop.

The only way in which cells can make themselves independent of the temperature of the surrounding environment is by forming communities which are large enough to include a self-made climate, and then developing mechanisms for stabilising that climate to the advantage of all the constituent members. The mere growth in size which such a congregation produces goes some way to protecting the organism from fluctuations in temperature: the bulkier an object is, the longer it takes to lose heat to a cold environment or to gain it in a hot one. But unless a creature can *actively* stabilise its own temperature, its cells will eventually bear the brunt of what is happening in the outside world. This is what happens to amphibians and reptiles, whose body temperature goes up and down with the weather. One of the reasons why mammals are so adventurous and versatile is that their body temperatures are independent of their surroundings. When it is chilled, the mammalian skin automatically insulates itself to prevent valuable heat from being lost: the superficial blood vessels constrict and divert warm blood away from the cold surface; in animals with fur, the pelt bristles, trapping an insulating blanket of air (man shows a futile vestige of this by coming out in goose pimples); if the body temperature continues to fall, the nervous system automatically switches on shivering, which, though mechanically useless, generates enough heat to compensate for the abnormal loss. Conversely, when body temperature begins to rise, the vessels of the skin automatically dilate, so that the blood can radiate its surplus heat through the body surface; this process is reinforced by sweating, which chills the skin as the moisture evaporates.

The efficiency of living cells also depends on the composition of the fluids by which they are surrounded. Single-celled animals can counteract small fluctuations in these, but there is a limit to their powers of adjustment. Multi-cellular animals are much more resilient: they can enclose and defend an inland sea whose chemical composition is supervised and regulated by the kidney. When surplus water lowers the

On page 132 *An inanimate object is unable to maintain its temperature. In this thermographic sequence a kettle filled with boiling water gradually loses heat into the surrounding atmosphere and eventually becomes thermally invisible.*

On page 133 *When the author stepped out of a hot bath his skin cooled but since he was, and still is, a living body he remained thermally distinct from his environment.*

Claude Bernard, 1813–78, the founding father of experimental physiology. He crowned the pioneering achievements of Bichat and Magendie, demonstrating the processes of self-regulation through which the living creature attained the stability of its internal environment.

The American physiologist, Walter Cannon, developed Claude Bernard's notions of automatic regulation and established the general principle of homeostasis — the processes by which the body maintains the most favourable conditions for its own function. Economists had already recognised this tendency in the behaviour of the market. When Cannon and his colleagues demonstrated the homeostatic principle in living organisms the idea was taken up with renewed interest by social theorists. Like blood, metaphors circulate through the intellectual community.

concentration of the body fluids, the kidney automatically lets out an extra volume of dilute urine; when water is in short supply, the urine becomes scanty and concentrated. The system exercises differential control over the retention and release of mineral salts as well: it can hold back sodium and eliminate surplus potassium. By orchestrating these and many other reactions, the kidney creates a favourable constancy for the living cells of the body.

Paradoxically, the communities which isolate living cells from the dangerous variations of a fickle environment become so secluded that they create problems of their own. The cells which comprise such an enclave can no longer disperse the by-products of their metabolism into the surrounding world. Like towns clustered along the shore of the Mediterranean, they empty the toxic by-products of their activity into the inland waters of the body itself, and, unless there are devices which can detect and void these substances, they begin to accumulate and pollute the internal environment. As a creature becomes larger and more complicated, these self-inflicted dangers begin to outweigh those imposed by its immediate surroundings. One of the most important functions of the kidney is the removal of substances created by the body itself: when the kidney begins to fail, as a result of infection or severe haemorrhage, the most immediate risk to the life of the patient is the accumulation of his own toxic substances.

The regulation of temperature and chemical composition are two of the ways in which organisms create what the nineteenth-century French physiologist Claude Bernard called 'un milieu intérieur' — a domain of physiological tranquillity without which the life of the individual cells would be nasty, brutish and short. Creatures which can create and defend such a domain have established a portable environment for the cells of which they are made, enabling them to exploit situations where the unprotected cells would be overwhelmed. The polar bear is able to colonise the Arctic because its brain cells are enclosed in a self-regulated tropics. As Bernard pointed out, the constancy of the internal environment is 'the absolute condition for a free life'.

Once Bernard had formulated the concept of the *milieu intérieur*, physiologists became interested in the mechanisms which defended and regulated its liberating constancy. It soon

135

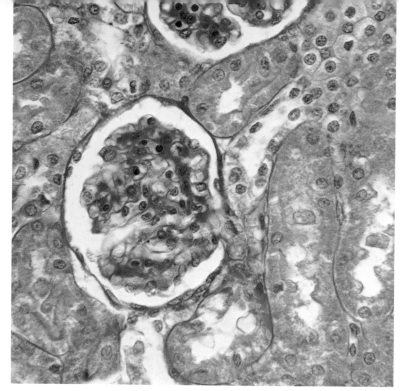

The filtration unit of a human kidney passing through this knotted blob of capillaries. The blood is filtered under high pressure. As it passes through the labyrinth of the renal tubules the filtrate is selectively processed in order to preserve the mineral constitution of the internal environment. The selection and reabsorption is under the control of various hormones, the most important of which are the steroids secreted by the adrenal gland.

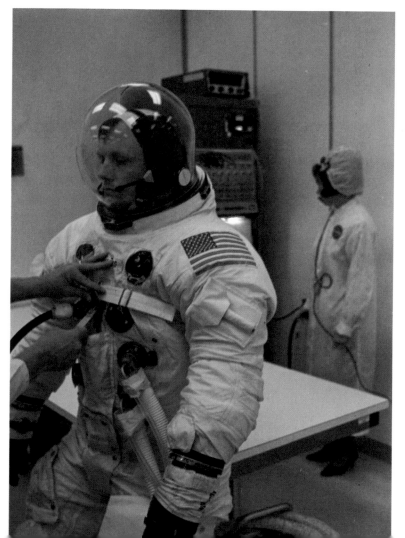

The homeostatic mechanisms of the healthy body preserve a constant internal environment in the face of wide variations in the outside world. But there is a limit to the physiological powers of compensation and if the external environment varies beyond what is physiologically expected it may be necessary to provide the subject with a portable version of his normal surroundings.

Nowadays we can manipulate the mineral composition of the internal environment, differentially supplying all recognised deficits. The management of fluids and electrolytes represents one of the greatest, if not the greatest, advances in medical practice.

became apparent that these homeostatic mechanisms exemplified a common principle, which is nowadays referred to as 'negative feedback'. In this process, some of the energy of a working system is used to counteract unacceptable variations in its own behaviour. If the system begins to drift away from some favoured norm — if it becomes too hot, runs too quickly, if its chemistry becomes unbalanced — the departure is detected and a measured instruction is fed back into the working mechanism, introducing the appropriate adjustment.

This principle was applied to technology long before it was recognised in the human body. Even in antiquity, engineers saw the advantages of machines which could regulate their own behaviour, but the mechanisms remained feeble and ineffectual until the 1780s, when English millwrights introduced a device which successfully stabilised the speed of windmills. The main axle was geared to a mechanism which would automatically refurl the canvas on the sails if they began to rotate too fast. This reduced the area exposed to the wind, and the windmill began to slow down. A few years later, the Scottish inventor James Watt abstracted the principle and saw that the steam engine, like the mill, could use some of its own motive power to compensate for unexpected variations in the boiler. A pair of spherical weights was slung from sym-

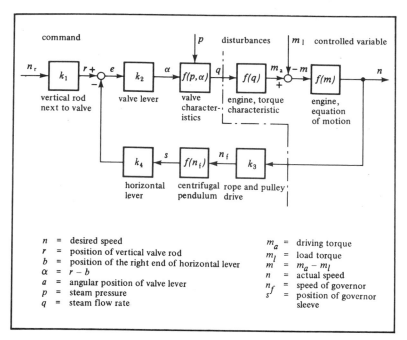

Mathematical block diagram illustrating the fundamental principle of negative feed-back as exemplified by Watt's centrifugal governor.

The gearing mechanism through which an engine lends some of its own motive power to the governor (opposite) which regulates its action.

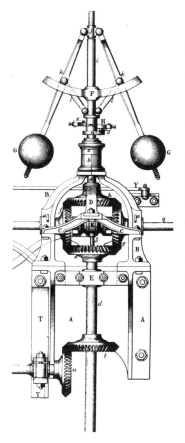

metrically hinged levers attached to a vertical shaft, which was made to revolve by means of cogs geared to the main axle. The faster the engine went, the wider the weights swung out under centrifugal force, pulling up a metal sleeve which was linked to the inlet valve of the boiler. This reduced the input of steam and lowered the power of the boiler, and the speed of the main axle began to diminish. Now spinning more slowly, the metal spheres fell back under their own weight and drew the metal sleeve down with them. The inlet valve reopened, and the engine started to re-accelerate. By hunting between these two extremes, the rotation of the driving axle eventually settled down at the speed which the engineer regarded as the most efficient.

During the nineteenth century and after, engineers and mathematicians realised that this principle could be applied to any system whose efficiency depended on the maintenance of a steady state, and it was only a matter of time before biologists recognised that the living organism regulated itself in the same way: it was a republic of servo-mechanisms whose collective action realised Bernard's homeostatic ideals.

This is not to say that, in the absence of any recognisable threats to its efficiency and well-being, a creature could pur-

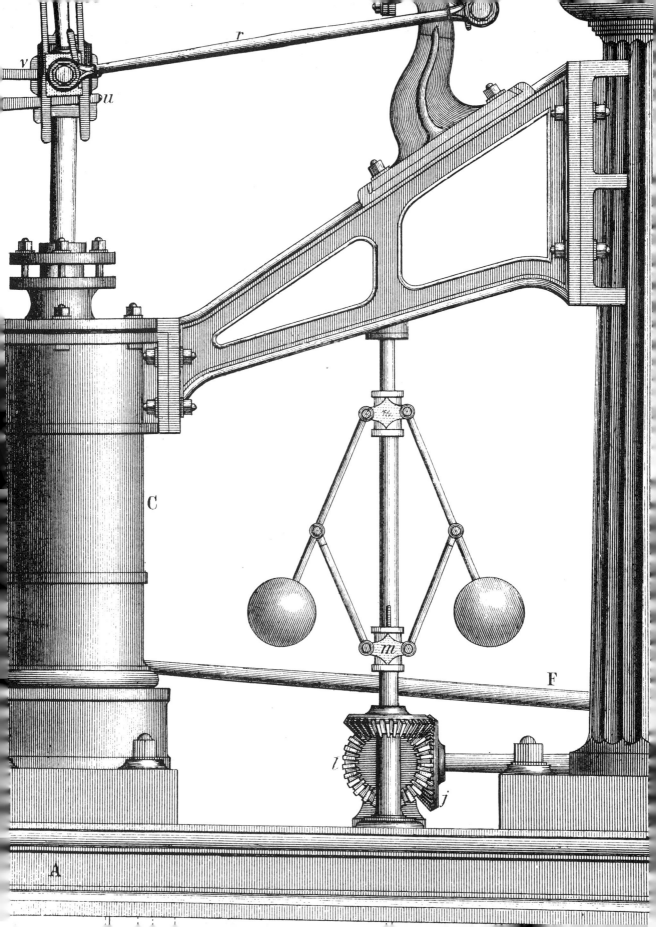

sue the main business of life undistracted by the work of maintenance and self-preservation. The nature of the physical universe is such that the mere existence of a living organism, the mere fact that it is distinguishable from its environment, means that it is in a state of jeopardy. By the middle of the nineteenth century physicists were forced to acknowledge that the physical universe tends towards a state of uniform disorder, a levelling down of all observable differences, and that left to themselves things will cool, fall, slow down, crumble and disperse.

In such a world the survival of form depends on one of two principles: the intrinsic stability of the materials from which the object is made, or the energetic replenishment and re-organisation of the material which is constantly flowing through it. The substances from which a marble statue is made are stably bonded together, so that the object retains not only its shape but its original material. The configuration of a fountain, on the other hand, is intrinsically unstable, and it can retain its shape only by endlessly renewing the material which constitutes it; that is, by organising and imposing structure on the unremitting flow of its own substance. Statues preserve their shapes; fountains perform and re-perform theirs.

The persistence of a living organism is an achievement of the same order as that of a fountain. The material from which such an object is made is constitutionally unstable; it can maintain its configuration only by flowing through a system which is capable of reorganising and renewing that configuration from one moment to the next. But the engine which keeps a fountain aloft exists independently of the watery form for which it is responsible, whereas the engine which supports and maintains the form of a living organism is an inherent part of its characteristic structure.

The fact that the mechanisms responsible for maintaining life are virtually indistinguishable from the structures they support is one of the reasons why it took so long to identify their existence. Even primitive biologists knew that the maintenance of life was a strenuous labour, but in the ancient world work was invariably performed by laborious devices, so that when human beings first began to speculate about their own characteristic 'go', they understandably sought the expla-

Two paradigms of permanence. Statues which 'retain' their form. Fountains which 'enact' it. The marble which the sculptor used remains as long as the statue survives. The water which forms the fountain is always renewing itself.

nation in the most unremittingly strenuous parts of the living body: those organs that seemed to go on their own, those physiological actions whose very spontaneity suggested that they were the prime movers of the living process. For more than 2,000 years, the heart, blood and lungs were regarded as the principle agents of life. Modern biology came into existence only with the recognition that the vital impetus was distributed throughout the living tissues of the body, and that the heart, lungs and blood, far from being responsible for life, were kept alive by biochemical processes which they shared with all the other structures of the living body.

(Opposite Michelangelo, statue of David, Florence. Below Fountain in the gardens, Villa d'Este, Tivoli)

4 · The Breath of Life

THE UNREMITTING REGULARITY OF BREATHING MAKES itself obvious to everyone who performs it. Although one usually breathes without a second thought, it is possible to do it consciously — in which case our attention is at once drawn to it whenever its rhythm is interrupted or obstructed. But although one is painfully aware that one is living on borrowed time as long as one is not breathing, it is not immediately apparent what one is denying oneself by abstaining from it. Eating and drinking supply the body with something that has an obvious existence over and above the efforts used to obtain it, but the commodity supplied by respiration is entirely invisible and almost intangible. A careless observer could easily get the impression that what was satisfying about breathing was the muscular exercise and nothing more.

But since the existence of air betrays itself in the form of winds and draughts, blown leaves and ruffled water, and since the act of breathing reproduces these effects on a small scale, one could at least infer that by breathing in one is replenishing oneself with the invisible substance that stirs the outside world. Once men recognised the existence of air or breath, the very fact that its presence was so hard to detect endowed it with a magical fascination. That life could be so easily extinguished for want of something so thin and meagre convinced ancient scientists that they had identified the animating principle of the universe. For the Pre-Socratic philosopher Anaximenes (active around 550 BC), air, breath or pneuma was the sovereign element of the cosmos, the primal substance from which all others took their origin. By drawing breath, the living body was renewing its vitality from moment to moment, replenishing its existence and its spontaneity by drinking in the material which bound the cosmos together.

Anaximenes' proposal does not really count as a scientific theory. He was simply re-stating the problem of respiration in colourful but uninformative terms. And by expressing himself in this way, and asserting nothing that was not already known, he anaesthetised curiosity and forestalled the possibility of future research. It is not that his suggestion is untrue, but that it hasn't even got the interesting possibility of being false. No conceivable evidence would count for or against the claim, because all that it says is that life is maintained by the intake of a principle which maintains life. It throws no light

On the previous pages
Air on the move.

Gillray, 'Windy Weather'.

WINDY WEATHER.

Opposite   *The first breath.*

144

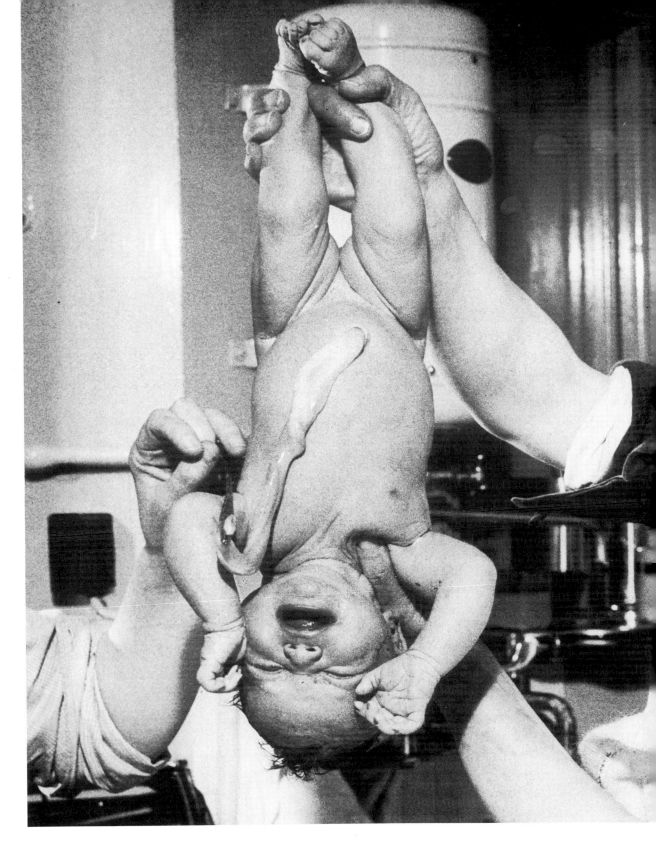

on the character of life itself and gives no insight into the physiological usefulness of air.

By the fourth century BC the vague, unprofitable archetype of air as the source of life had given way to a much more specific model, one which assigned a function to the intake of breath. Instead of asserting that air was the abstract principle of vitality, Plato and Aristotle identified its role in a physiological process. According to them, the maintenance of life depended on the existence of an innate physiological heat, a productive warmth without which the body would be unable to assimilate food, move its limbs or reproduce itself. Aristotle assumed that this profitable flame burnt within the cavities of the heart, and that unless it was periodically cooled by a ventilating current of air it would consume itself and go out. This theory was given its conclusive form by the Roman physician Galen, who bequeathed to the world a canon of physiological ideas which provided the framework of all biological thought until the birth of the modern scientific era in the sixteenth century. For Galen, the fire in the heart was not simply an innate principle but part of a systematic process. It was the agent of chemical transformation, brewing, cooking, distilling or smelting solid food to produce the spiritual substance responsible for movement and sensation. Breath not only fanned this process, it provided one of its essential ingredients. The air interacted with the blood in the heart, catalysed by the subtle incandescence of the vital heat.

Although this picture now strikes us as bizarre, it compares favourably with the one painted by Anaximenes because it has the exciting possibility of being wrong. It replaces abstract principles with material processes, and, like all useful metaphors, brings into the same domain of thought two processes which previously seemed to have nothing in common — the warmth of the blood and the heat of a lamp, the ventilation of the chest and the smelter's bellows. The very improbability of such an analogy gives speculative thought a useful purchase. Unlike Anaximenes' proposal, it invites experiment. If the heart is a lamp or a furnace, it should be hotter than the other parts of the body — and a simple experiment would show this. If breath cools and refines the process, there must be a free airway between the lungs and the ventricles. Does such a passage exist? Is air to be found in the blood vessels

The smelter's furnace provided a persuasive metaphor for explaining the transformation of venous into arterial blood. (Right   Greek vase, British Museum.
Below   *Athenian red-figure Kylix from Vulci, Antikenmuseum, Berlin)*

linking the heart and the lungs? By suggesting an arrangement whose parts are systematically related to one another, by comparing it to a known mechanism, where 'this' happens followed by 'that', a programme of practical research is created, a tissue of soluble problems.

The scientists of antiquity did not pursue the matter along these lines — not, as is sometimes claimed, because they preferred theory to experiment, but because they did not yet appreciate the critical *purpose* of experiment. A modern scientist expects a theory to have a set of falsifiable implications: he knows that he has a duty to subject these to the most drastic tests he can devise, and that a good theory will automatically suggest the experiments which would lead to its overthrow. One of the reasons that ancient scientists did not systematically check the implications of their theories was the fatal dissociation between the speculative professions and the technical ones. Those who devoted their life to thought disdained people who worked with their hands. Aristotle, for instance,

147

excluded artisans and manufacturers from any serious role in statecraft, insisting that the exclusive concern of the free man was the manipulation of abstract ideas and that only a slave should concern himself with technical matters.

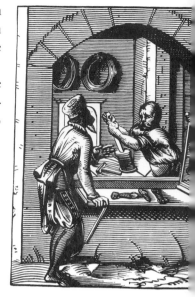

This attitude inflicted almost irreparable damage on the structure of scientific thought by underestimating the cognitive value of technical craftsmanship. What Aristotle failed to see is that making something, persuading something to work, is a form of knowledge in itself, presupposing ideas and generating still more. The craftsman who handles physical materials in order to make a go of some project familiarises himself with their properties and soon discovers what they will and what they won't do, and although his theoretical notions are not as explicit or abstract as those of a speculative philosopher, the fact that they are directed towards a practical end means that they are bound to be submitted to the harsh test of experiment. The craftsman and the manufacturer can't afford to theorise in a vacuum and have no time to tinker with materials in the hope of finding something interesting or illuminating. For a stonemason or a metal worker, for a glassblower or a mining engineer, all knowledge is the pursuit of the feasible.

If early theoretical scientists had identified themselves more closely with such ingenuity, they would automatically have recognised the advantages of submitting their proposals to the test of practical experiment. By familiarising themselves with the engines, tools and mechanisms which technologists invented in order to make their tasks easier, they would have learnt to generalise some of the mechanical principles which hold the physical world together. Pulleys, levers, pumps and presses not only do the tasks for which they are designed: they also exhibit fundamental principles of physical action — they are exemplary and almost diagrammatic displays of natural processes. By divorcing themselves from experience of this sort, theoretical scientists seriously disabled their imaginations, and, as a result, scientific thought remained crippled and inert for more than 1,400 years after the birth of Christ.

It wasn't until the beginning of the sixteenth century that a spirit of profitable reconciliation began to emerge. In the astounding cupola which he succeeded in raising over the

Francis Bacon was one of several sixteenth-century writers who recognised that science had disabled itself by overlooking the work of craftsmen and technicians. 'For twice a thousand years the sciences stood where they did and now remain almost in the same condition . . . whereas in the mechanical arts, which are founded on nature and the light of experience we see the contrary happening, for these are continually thriving and growing . . . at first rude, then convenient, afterwards adorned, but at all times advancing.'

Santa Maria del Fiore in Florence, Brunelleschi demonstrated the unprecedented advantage of marrying the practical technology of the stonemason with the abstract mathematics of classical antiquity, and by the end of the century argumentative craftsmen were beginning to publish books insisting that the practical experience gained from pottery, navigation, metallurgy and mining would inevitably yield more truths about nature than all the traditional dogmas of antiquity. In England, Francis Bacon was one of the first to insist that the snobbish disregard for manual labour and technical skill had paralysed the pursuit of useful knowledge and that much was to be learnt from those whose daily work was taken up with bending nature to the service of man. He advised his contemporaries to study the work of smiths and smelters, soapboilers, engineers and dyers. Previously, this advice would have fallen on deaf ears, but by the start of the seventeenth century, the rise of Puritanism had given intellectuals a new respect for work and labour, and a positive admiration for the use of practical intelligence.

Although the influence of Puritanism can be overemphasised — the revival of ancient learning was just as important as the critical impatience directed against it — it is interesting to note that a disproportionate number of those who came together to study experimental science in London and in Oxford belonged to the Puritan faction. Taking their inspiration from Bacon and the great Czech educational reformer Comenius, men such as Samuel Hartlib and John Dury established committees for the advancement of learning, recommending that academic scholarship should be reinforced by the practical wisdom of trade, craft, medicine and agriculture. Henceforth, the whole 'theatre of nature' was to be systematically ransacked for the improvement of man and the greater glory of God.

At the end of the Civil War, a group of remarkable individuals who had been meeting in London during the 1640s incorporated themselves into a society, the express purpose of which was to confront nature with a list of soluble questions and, where necessary, put her to the torture of direct experiment. Shortly after the restoration of Charles II, the society obtained an official charter, and became the Royal Society of London. The first problem to which the members addressed themselves was the nature of air and the part which it played in both breathing and combustion.

Although the ancient Greeks had recognised the existence of air, maintaining that it was one of the four fundamental elements, they had paid no attention to its material properties, apart from suggesting that it was by definition warm and wet.

In 1643 Torricelli showed that a column of mercury was supported by the weight of atmospheric air and not, as his predecessors supposed, by Nature's horror of a vacuum. He observed that the column of mercury varied from day to day. 'Nature,' he suggested, 'would not, like a flirtatious girl, have a different horror from one day to the next.'

Brunelleschi's great dome on the Santa Maria del Fiore, Florence.

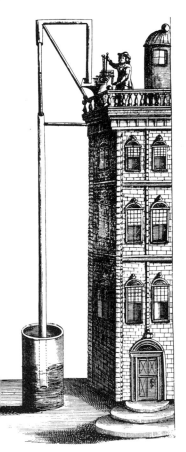

A variation of Torricelli's famous barometric experiment. From experiments such as these it became apparent that Nature's famous abhorrence of a vacuum varied according to the weather, the altitude and the density of the liquid. It became clear that the column of liquid was supported *by the positive pressure of the atmosphere rather than being* suspended *by the vacuum at the top of the tube.*

The Italian Torricelli demonstrated, however, that in spite of its invisibility it had weight, and that it was this which supported a column of fluid in a barometer tube — a fact confirmed by Pascal in a classical experiment: he took a barometer to the top of the Puy-de-Dôme and showed that the height of the column diminished as the air thinned out.

These experiments excited considerable interest throughout Europe, and the development of pumps made it possible for scientists to manipulate a substance which had previously flowed elusively through their fingers. In England the Irish aristocrat Robert Boyle took especial interest in this device and with his own pneumatical engine created a vacuum within which life and flame were extinguished. Both breathing and combustion, it seemed, were dependent upon the same invisible substance — the very material which supported Torricelli's column of fluid.

This was the first experimental indication that breathing was more than a thoracic exercise and that the breathing animal was indeed supplying itself with physical substance. It was not a conclusive proof, however. The next essential step was taken by Boyle's Royal Society colleague Robert Hooke, who had been appointed curator of experiments in 1664. On 24 October 1667, Hooke performed a memorable experiment which established that respiratory movements were simply a way of filling the lungs with air. Opening the chest wall of a dog so that it could no longer exercise its ribs, he showed that the animal could be kept alive as long as fresh air was artificially blown through its lungs, and that, as long as the air was artificially replenished by means of a pump, the chest could be immobilised without harming the animal at all.

Why did the lungs depend on this supply of air? What possible function could this meagre substance have? According to the traditional theory, it was cooling and ventilating the fire in the heart. But by the end of the sixteenth century this particular theory had been subjected to severe criticism. It was already known, for example, that it was impossible to pump air down the windpipe and into the heart itself, and that the blood vessels connecting the heart and lungs were filled only with blood. The Italian physiologist Borelli repeated these experiments and concluded that breathing had no part in ventilating the vital flame in the heart. He realised, however, that 'so great a machinery of vessels and organs of the lungs must have been instituted for some grand purpose': that the lungs were not simply a throughway carrying air from the atmosphere into the heart, but a channel within which the blood came into contact with the air. Until the sixteenth century this suggestion would have made very little sense: no one had suspected the existence of a pulmonary circulation. Towards the middle of the century, however, physiologists began to suspect from the size of the blood vessels going to and from the lungs that there was a substantial flow of blood through the lungs and that this was the only way in which the blood entering from the right side of the heart could reach the major exits on the left-hand side. It became apparent that, far from the air entering the lungs in order to gain access to the heart, the blood was flowing through the lungs in order to gain access to the air.

What was the result of this encounter? What change did the blood undergo as a result of such exposure? To answer these questions, another English physiologist, Richard Lower, exploited Hooke's technique of artificial respiration. It had been known since antiquity that blood was purple as it entered the heart and bright red leaving it, but it was thought that this colour change took place within the heart itself, as the result of the blood being brought into direct contact with inspired air. Lower showed that the transformation took place as the result of the blood's passage through the lungs. He opened the chest of a dog and slit the vessel through which blood left the lungs. As long as the artificial respiration was maintained, the blood running out of the lung remained bright red, but as soon as the flow of air was stopped the blood went purple.

Lower was not satisfied with this simple experiment. To prove that the colour change was the result of exposure to the air and not merely of a transit through the lungs, he poured some blood into a long glass vessel and found, as he had expected, that the clot turned bright red on the upper surface where it was in contact with the air and remained dark purple in the depths. When the clot was turned upside-down, the pattern was slowly reversed: the purple blood reddened when exposed to the air whilst the red blood, now buried in the depths of the glass, went purple. The conclusion was inescapable. The blood changed its colour as a result of being exposed

Von Guericke's experiment with a vacuum. Although he was unaware of Torricelli's work, von Guericke showed that air could exert considerable mechanical pressure. He fitted together two metal hemispheres and sucked the air out of the sealed sphere which resulted. The atmospheric pressure kept the two hemispheres together.

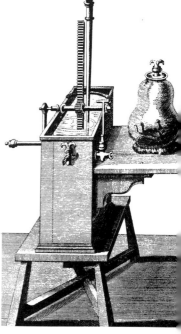

Boyle's experiments with 'pneumatical' engines extended and amplified the seventeenth-century interest in the weight, 'spring' and composition of air.

*Richard Lower, c. 1631–1691.
(School of Jacob Huysmans)*

*If a hole in the heart allows
venous input to 'shunt' from
right to left without having to
go through the North-west
Passage in the lungs, the
arteries will supply under-
oxygenated blood to the tissues,
and the tongue, lips and
fingernails will appear blue –
so-called cyanosis.*

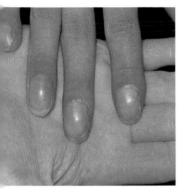

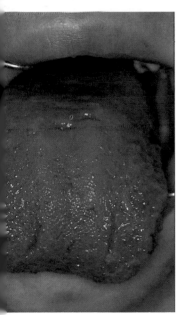

to fresh air, and although this exposure normally takes place in the ventilated lung, Lower's second experiment had proved that it was a straightforward chemical reaction between the blood and the air.

It seemed unlikely that the blood made its way through the lungs simply in order to enjoy the frivolous luxury of a colour change. Presumably, it underwent a physiologically useful transformation. Since Lower had shown that the blood of an asphyxiated animal remained purple, it seemed reasonable to assume that whatever it was in the air that turned the blood scarlet was the same thing that kept the animal alive, and that the colour change from purple to scarlet was the outward sign that the blood had removed some valuable substance from the air in the lung. Was this substance air itself, or was it some-thing contained in the air?

Up to this time, air had been regarded as an element and, as

153

such, the idea that it might be composed of different substances was inconceivable. The physiology of respiration could make no further progress until scientists recognised that the essential ingredient of air — the part which supported life and kept blood red — represented a comparatively small fraction of the whole. This important discovery was made by John Mayow, who was born in 1643. Although he qualified as a lawyer, Mayow spent most of his life studying chemistry and physiology and, by the time he died at the age of thirty-nine, he had discovered oxygen in all but name.

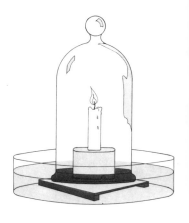

In the first of his great papers, Mayow reminded his readers of how Boyle had extinguished life and flame by sucking air out of a sealed glass vessel, but he went on to show that only a certain proportion of the air was capable of supporting respiration and combustion. He placed a candle on a raised platform in a bowl of water and lowered a bell jar over the top. His argument was that if the candle consumed something in the air, the water level would rise as the atmospheric pressure fell inside the vessel. He noticed, however, that the candle

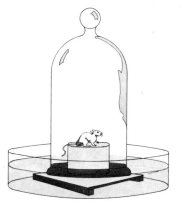

John Mayow, 1640–79. Mayow's chemical investigations hardly bear comparison with Lavoisier's majestic research, but his work foreshadowed that of his successor and if his conclusions had not been swamped by the phlogiston theory, the discovery of oxygen might have been made a hundred years earlier than it was.

154

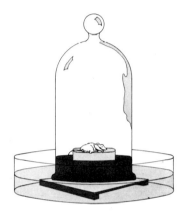

went out by the time the water level had risen no more than an inch or so, suggesting that, although there was plenty of air left in the vessel, what remained was useless for combustion. He repeated the experiment with a mouse: again, the animal expired long before the water level had risen to fill the vessel. Obviously, only a small proportion of atmospheric air was useful for burning and breathing, and the rest was incapable of supporting either life or flame. One question remained: were the candle and the mouse both using the same substance, or did the air contain *two* active ingredients — one which supported life and another which maintained flame? Mayow introduced a mouse and a candle into the same vessel and found that each expired sooner than either would have done if left on its own. This implied that both were competing for the same ingredient, and that in any given volume of air this ingredient represented no more than one-fifth of the total.

The next step was to identify what the valuable substance was, and Mayow went on to suggest that it was similar to, if not actually identical with, something contained in so-called potash of nitre, an essential component of gunpowder. His argument went as follows: a combustible material like sulphur will burn only if it is plentifully supplied with air; but if it is mixed with potash of nitre, it will burn vigorously, quite explosively in fact, even in a vacuum; Mayow inferred that potash of nitre, or saltpetre as we now call it, must contain or be partly composed of the very substance which made *air* capable of supporting combustion. For this reason, he called the active ingredient 'nitro-aerial spirit', and suggested that it was identical to the substance for which both candle and mouse were competing as they expired in the sealed vessel.

What makes these experiments so recognisably modern is the tight sequence of hypothesis, deduction and experiment, and, above all, the abstract concern for ratio or quantity. The productivity of such a method speaks for itself: to all intents and purposes, Mayow anticipated the French chemist Lavoisier's discovery of oxygen by more than 100 years. Unfortunately, the significance of his discovery was lost on his contemporaries: scientists were seduced by an alternative theory of combustion which, superficially at least, gave a much more plausible account of the process of burning.

When wood or wax burn, the immediate impression is one

Lavoisier did for chemistry what Newton had achieved for physics. He penetrated the deceptive surface of qualitative appearances and recognized the underlying structure of quantitative relationships. It was this that enabled him to see the resemblance between respiration and combustion. (David, 'Lavoisier and his Wife', Metropolitan Museum of Art, New York)

of destruction and loss. The candle vanishes; the wood is reduced to a light ash. The obvious deduction is that such materials owe their combustibility to a volatile substance which vanishes during combustion. Lavoisier's experiments show how dangerous it is to accept the immediate evidence of the senses and that the apparent loss of weight gives a misleading impression of what is actually happening when a substance burns. If the observations are confined to the visible products of combustion — to the ash alone — the loss of weight is undeniable. But if the less noticeable products — the water vapour and the carbon dioxide — are taken into consideration it is apparent that, far from losing something to the atmosphere, the combustible material combines with something in it.

Lavoisier identified this substance with the gaseous material which the English chemist Joseph Priestley had succeeded in generating from the ash or calx of certain heavy metals which had previously been heated in air. It was the same substance Mayow had already obtained from saltpetre. Lavoisier proved that it was also responsible for the support of life, and that breathing is therefore analogous if not identical to burning:

> We can state in general, that respiration is but a slow combustion of carbon and hydrogen, similar in all points to that taking place in a lamp or a burning candle and that from this point of view animals which breathe are really combustible bodies which burn and are consumed. In respiration as in combustion it is the atmospheric air which supplies the oxygen, but in respiration it is the substance of the animal itself, the blood, that provides the combustible material.

On the face of it, Mayow and Lavoisier's theory seemed to revive the ancient idea of the biological furnace, with the obvious difference that the combustion was now localised in the lungs rather than in the heart. This suggestion, however, raised serious problems, for by Lavoisier's time thermometers were sensitive enough to show that the blood in the lungs was no hotter than it was in any other organ of the body.

By the middle of the nineteenth century, it had gradually become clear that the physiological combustion was distributed throughout the living substance of the body, and as the century drew to a close scientists were beginning to realise that the generation of heat, far from being the source of life, was an incidental feature of a much more fundamental process: the mobilisation of biologically useful energy — the energy needed for growth, movement, secretion and nervous activity; the energy, in fact, without which the living cell would be unable to maintain its volatile configuration.

But before this could be appreciated, scientists had to recognise the concept of energy as such: to realise that the only way of resisting or counteracting the tendency of the universe to move towards a uniform state of dilapidation was by active exertion, and that what was consumed in such exertion could be neither created nor destroyed but only transformed or re-allocated.

This concept, like so many others in the history of physiology, emerged largely as the result of the attempts which were made to improve the performance of a man-made instrument, in this case the steam engine. By eliminating unproductive heat losses and reducing the friction of the moving parts, the amount of work which could be got out of a steam engine could be made to approach the amount which had to be done to make it operate. To approach but never equal: there was a limit, it seemed, to the improvement of the performance of such machines, and theoretical engineers had to acknowledge that even under the most favourable circumstances imaginable — an ideal world where friction was non-existent and insulation was perfect — anything which performed or worked would continue to do so only if it was supplied with energy; that is to say, if some equivalent work was performed on its system.

Energy, it was clear, was not a commodity which existed independently, but a unit of mathematical currency that enabled scientists to measure the working transactions of the physical universe. When nineteenth-century scientists said that energy existed in a given system or that it was locked up in a particular substance, they were not implying, as the ancient Greeks might have done, the existence of some elusive, imponderable principle; they were merely predicting how much alternative work could be obtained from the substance or system. The steam engine had demonstrated that heat could be converted into mechanical work, and in the 1840s the English physicist Joule showed that if motive power was supplied to a system, heat could be generated from it in a consistently quantifiable fashion. He designed a system of brass paddles, which he revolved in a carefully insulated water bath, and found that the quantity of heat produced by the friction was always proportional to the amount of mechanical force which had been expended. By 1850, scientists such as Helmholtz were aware that the universe was a closed system, transforming and redistributing energy in readily measurable types of work.

As chemistry developed in the latter half of the nineteenth century, it was gradually recognised that energy was stored in the chemical bonds which held molecules together, and that when these bonds were broken, as they were in combustion for example, the energy appeared in more noticeable forms — as heat, for instance, or light. Until the sub-microscopic chemistry of the living cell was elucidated, however, it was impossible to see how the living organism gained access to the energy which was locked up in molecular bonds. Lavoisier's revolutionary intuition that living and burning were closely related to each other was useful as far as it went, but since there was no recognisable fire in the tissues, it was difficult to see what the organic counterpart of combustion really was.

This problem was gradually solved by work which the French chemist Pasteur began in the 1860s and the German chemist Büchner developed in the 1880s. In the course of research done on behalf of the French wine industry, Pasteur discovered the part micro-organisms play in degrading sugar into alcohol. While investigating certain faults in fermentation, he found that the different species of micro-organisms did not always break sugar down into the same substances, but that each particular species produced its own characteristic spectrum of products, such as acetic acid and butyric acid. From this Pasteur deduced that fermentation was the way in which living organisms nourished themselves. Here, it seemed, was the counterpart of Lavoisier's chemical combustion, and when Büchner demonstrated that fermentation could continue in the absence of cells — after they had been ground up, filtered and the extract added to sugar solution — it became apparent that the cell was not a

Joule's apparatus for measuring the mechanical equivalent of heat. With the use of apparatus such as this Nature was made to yield the secrets of her economic transactions. In the new age of capitalism it represented a vivid illustration of the labour theory of value. Joule's work coincides with some of Karl Marx's early manuscripts.

159

simple bag of protoplasm but an organised chemical factory with components specially geared to break down foodstuffs into smaller and smaller molecules.

With the spectacular development of biochemistry during the past fifty years, we now know that the cell comprises an elaborate chemical scaffold along which is strung a series of intermediate pathways, and that along these pathways materials are systematically broken down in successive stages, with the aid of oxygen supplied by respiration. In the course of this process chemical bonds are sundered and the energy which is released is handed on to a master substance which is found in every living cell, from yeast to neurones.

This master substance, adenosine-tri-phosphate, known as ATP, which is synthesised at the expense of the systematic breakdown of sugar, contains three superlatively energetic chemical bonds, the breaking of which makes energy available in any one of the forms — mechanical work, electrical impulses, the energy needed for cell-division, secretion and synthesis — that might prove useful for the continuing life of the organism. By utilising oxygen and carbohydrate, a cell maintains its ATP credit — a biological deposit of general currency which is immediately cashable into whatever form of energy is called for at the moment. By continuously cashing its liquid assets of ATP, the cell also defends itself against the inexorable tendency towards its own dilapidation.

By taking in oxygen the organism is, therefore, not — as Anaximenes supposed — drinking in the mysterious principle of life itself: it is providing itself with the chemical key to unlock the resources of life which are distributed in and constituted by its own biochemical organisation. Oxygen is

ATP is the universal currency of biological energy. The wavy lines represent high energy phosph bonds. Each one releases 7,500 calories when broken. This energy can be made available to any of the functions of the living cell.

the indispensable combination with which the cell opens its own Swiss bank-account. For the modern biologist, life is no longer an assembly of inert substances huddled around a mysterious source of vitality, but an arrangement of matter whose collective transactions result in vitality. Like a fountain, the living organism retains its improbable configuration by borrowing sources of energy from the world around it and by conferring and re-conferring organisation upon the matter which is ceaselessly flowing through it. And, in order to do this, to exploit the energy resources of the substances it borrows from the outside world, the cell must have oxygen.

In a small, single-celled animal, the intake of oxygen presents no problems. The volume of living protoplasm is so small that the creature's surface is large enough to absorb all the oxygen it needs: it does not have to make active efforts to obtain it. But when an animal grows beyond a certain size, when it becomes multi-cellular, the volume of protoplasm

Below and overleaf   The delicate gills of the fan worm increase the respiratory surface without making the organs unduly cumbersome.

demanding oxygen outstrips the surface area which can provide it. It is a purely mathematical misfortune that as the volume of a solid body increases its surface area does not expand at the same rate. Consequently, as an animal grows larger, it runs the risk of being asphyxiated by its own bulk.

This means that there has to be an additional surface with the specific purpose of supplying all the oxygen such a large creature requires. Since this surface has to be many times greater than the overall area of the rest of the body, it poses an elaborate packaging problem: the absorptive area must be large enough to take in all the oxygen that is needed by a heavy, busy animal, but at the same time it must be pleated into a conveniently portable organ or gill. Each species has its own characteristic design. Marine worms are dressed overall with a frilly bunting of permeable fronds; molluscs and crustacea carry fans, cones and rosettes, the designs of which increase the surface without significantly adding to the creature's weight or volume.

These systems can sprout anywhere on the surface of the body. They may be grouped in a crown on the head, clustered along the back, or even around the anus. In fish, the gill system is associated with the mouth and throat: on each side of the pharynx, symmetrical pairs of corridors branch off at right-angles and open to the outside through gill-slits; by gulping and then compressing the closed mouth, the oxygen-filled water is pumped sideways through the corridors, each of which is lined with pleated membranes.

Pleats preserve the surface area while packing it into a smaller space.

In land animals the skin becomes even more impermeable than it is in fish, and since oxygen diffuses only across a damp surface, the respiratory membranes have to be buried within the body and kept in a moist atmosphere to prevent them from drying out. Gills dwindle, and lungs appear in their place. But the colonisation of land could not have begun unless there had already been substantial traces of an organ capable of serving as a lung. Such an organ made its appearance at a comparatively early stage in the evolution of fishes, as a blind outgrowth from the floor of the throat. In those types which remained in the ocean, this outgrowth lost its original connection with the throat and became an air-filled buoyancy tank or swim-bladder. But in those types which were beginning to flop around at the edges of the coastal mud, there

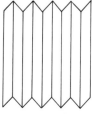

Australian lung fish. A rare survivor of the ancestral stock from which the land vertebrates took their origin.

The first division of the human bronchus, showing the entrances into the right and left lung, followed by further divisions of the bronchial tree.

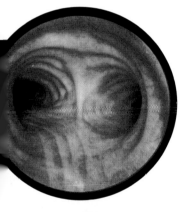

was an obvious advantage in retaining the ancient connection with the throat, since the tube could now serve as an airway. The lung-fish which still survive in Australia and South America show what these transitional forms might have been like. The barely functioning lung allows these types to outlive short periods of drought on the river banks, but, since they retain the original gill structure as well, they are not at a loss when the water level returns.

Such primitive lungs are little more than emergency devices; like the life-support equipment of astronauts, they allow short, adventurous excursions into a new environment. Before the vertebrates could permanently sever their connections with the water, the lung had to undergo three important modifications: its inside surface had to be enormously increased without enlarging the bulk of the organ; it had to be supplied with blood vessels so that the oxygen which was taken up could be swiftly distributed to the body at large; it had to be ventilated and systematically replenished with fresh air.

The first condition has been satisfied quite simply. As the primitive sac buds from the floor of the throat, it branches right and left. Each of these divisions forks again, and again,

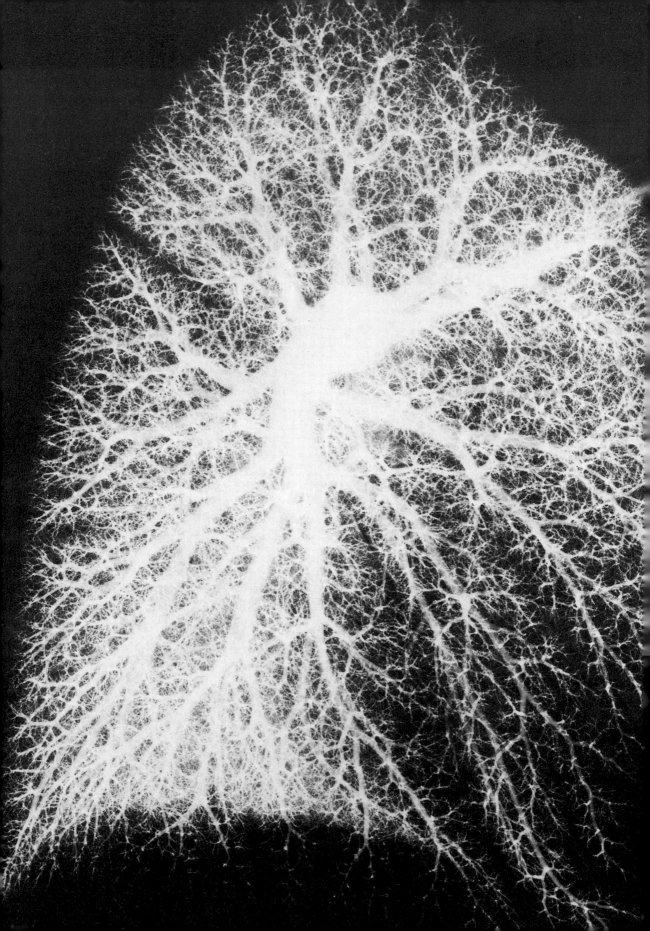

A busy inland harbour. Extra inlets from the main waterway increase the area for loading and unloading. These mercantile 'pools' correspond to the alveoli in the lung. (The Royal Docks, London)

An X-ray showing the myriad branches of the blood vessels in the human lung.

creating a vast tree of tubes. The tips of each branch end in millions of blind sacs, or alveoli, with walls so thin — not more than one cell thick — that gasses can diffuse easily.

When an animal is large enough to need lungs, however, its bulk is so great that the physical diffusion of oxygen will not be fast enough to satisfy the insatiable demands of its working tissues. There has to be an efficient transport system capable of carrying oxygen from the lungs to every part of the living body. The evolution of the respiratory system is therefore associated with the development of an energetic heart and rapaciously absorptive blood. These are the necessary conditions for the efficient provision of oxygen. One further device is also needed. Although the surface of the lung absorbs oxygen freely, the air which it contains soon becomes stagnant. In a motionless lung the air would soon become depleted of oxygen and at the same time prohibitively

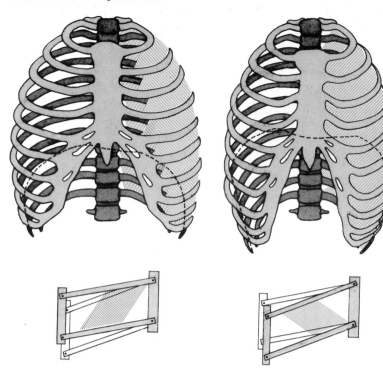

During inspiration the diaphragm, shown as a dotted line, descends, increasing the depth of the chest. In expiration it recoils to form a steep dome. The hinged movement of the ribs simultaneously alters the fore and aft diameter of the chest.

Microscopic section of the human lung specially injected to show the basket-work of capillaries.

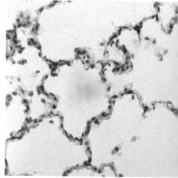

The thin-walled air-sacs or alveoli where the essential exchange of oxygen and carbon dioxide takes place. These are the respiratory wharves.

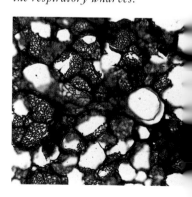

contaminated with carbon dioxide. The blood would be visiting a useless organ and, as Richard Lower discovered, would return to the circulation purple and vitiated. The lung must be actively ventilated so that the air in it is endlessly renewed.

The first scientist to appreciate the mechanics of ventilation was John Mayow. In 1670 he demonstrated with a mechanical model that air is drawn into the lung by enlarging the thoracic space which contains it. He inserted a bladder into the cavity of a pair of bellows, making the mouth of the bladder continuous with the nozzle. When the handles of the bellows were pulled apart, air rushed into the bladder, which emptied again as the bellows were squeezed shut.

But how are the thoracic bellows enlarged? If you trace the contour of the ribs from back to front, it is easy to see that they are slung from the backbone in a downward slant. When one breathes in, the muscles between the ribs contract, lifting them up like hoops on a barrel and, since the slant is thereby reduced, the diameter of the thorax is increased. At the same time the muscular diaphragm which forms the dome-shaped floor of the chest flattens, thus increasing the depth of the

The respiratory movements are driven by the rhythmic discharge of nerve cells in the brain stem. But the rate and amplitude of this unremitting broadcast can be altered by influences flowing in from many sources. These ensure that breathing serves the varying needs of the creature.

chest. As the size of the chest cavity increases, air rushes in to fill the lungs. To blow the air out again, the muscular effort simply ceases: the ribs drop passively into their original position, the diaphragm springs up into its dome, and the lung is squeezed.

This rhythmic alternation repeats itself throughout life, and although James Thurber had a cousin who put off sleeping as long as possible for fear he would quit breathing as soon as he lost consciousness, the rhythm is self-perpetuating and does not need conscious supervision. Like any activity which serves an unremitting need, breathing is dictated by an auto-

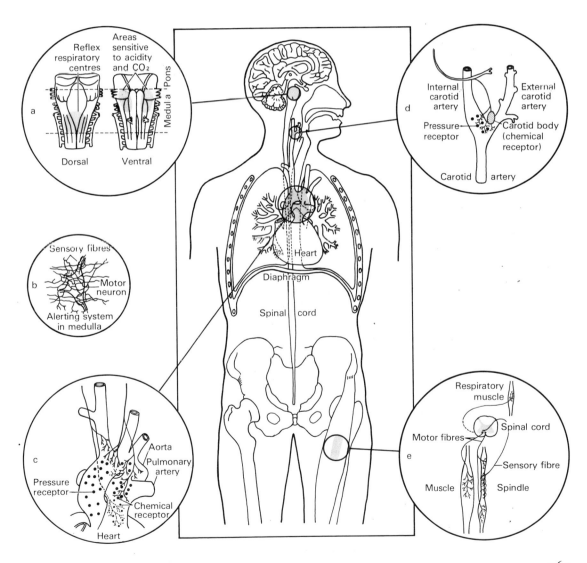

matic mechanism, so that it does not run the risks of absent-mindedness. But, unlike those of the heart, the muscles responsible for ventilating the lung are not spontaneously active. Strips of heart muscle will continue to beat even when they are isolated in a test-tube of warm saline: in fact, the individual cells can be seen to pulse when grown in tissue culture. The muscles of the thorax and diaphragm, however, have to be driven by a neurological motor which is situated in the brain stem just above the point where the spinal cord enters the skull. Under normal circumstances, this nerve centre broadcasts fifteen or sixteen inspiratory instructions per minute without having to be prompted or reminded by a conscious decision. The higher level of the brain can even be removed altogether without destroying this automatic life-support system, since the centre is driven by the inherent rhythmicity of the nerve cells which constitute it — rather like the battery-driven pulse of a Bulova watch. If this centre is destroyed, as it sometimes is by polio, or if it is disconnected from the thorax by a broken neck, the bellows become immobile and the patient has to be put on a mechanical respirator.

Although this nervous rhythm is self-perpetuating, it is within reach of conscious control. We cannot override the native rhythm beyond the loss of consciousness, but we can deliberately hold our breath, deepen it, or speed it up, and can borrow its activity in order to speak, sing, sigh, laugh, whistle, grunt and groan. As soon as these expressive interludes subside, the machinery reverts to the task of supporting life, but even in its humble, domestic role of keeping the lung regularly ventilated the respiratory centre shows enormous versatility.

As we saw in the previous chapter, any activity which serves a vital function automatically modifies its own behaviour in the light of its success or failure in serving that function. By feeding back the consequences of its own exertions, the activity ensures that they give satisfaction. The respiratory centre exhibits this principle with exemplary finesse. If the rate and depth of normal ventilation is enough to keep the cells of the body adequately supplied with oxygen and safely cleared of carbon dioxide, breathing inevitably stabilises itself at the level which preserves this satisfactory state of affairs. But if the oxygen concentration falls below this, or if the carbon dioxide begins to rise above it, the respiratory centre acts more energetically to restore the situation. Like the homeostatic devices considered in the previous chapter, the system is 'error-activated'.

All such systems have to have: (1) a detecting mechanism capable of registering unacceptable departures from a required state; (2) a communication system which can transmit signals whose size or intensity is proportional to the recorded error; and (3) a power system which can convert these signals into the actions needed to restore the *status quo*. In the case of respiration, the required state is a level of oxygen which allows each and every cell to obtain the energy needed for its own maintenance, plus a satisfactory freedom from the local accumulation of carbon dioxide. The variable power mechanism is the respiratory bellows. By increasing the rate or depth of breathing, oxygen is made more readily available and carbon dioxide more freely eliminated.

How is the error detected, and where? Each and every cell in the body is affected by local alterations of oxygen and carbon dioxide, but not every cell is endowed with the ability to set the necessary adjustments in motion. In a complex organism there is a noticeable division of labour: certain representative cells are set aside to act as respiratory monitors. Since the central nervous system is more susceptible to oxygen shortage than any other tissue of the body — it is the organ which shows the earliest sign of irreversible deterioration — the cells that supervise this vital factor are located in the brain itself. In fact, they are the very cells that are responsible for the maintenance of normal breathing. As it tastes the chemical composition of the blood circulating through it, the respiratory centre varies its own performance in order to stabilise its own immediate environment at the most favourable level. If the oxygen in the blood falls or the carbon dioxide level rises, the rate and amplitude of breathing automatically increases until the restoration of the required level eliminates the need for further action.

By acting for itself in this way, the brain also creates the most favourable conditions for the rest of the cells in the human body, thus making an indispensable contribution to the overall stability of Claude Bernard's *milieu intérieur*. In the 1920s physiologists began to recognise that such error-activated mechanisms were almost invariably duplicated by back-up systems working on comparable principles, and that the respiratory composition of the blood was supervised not only by nerve cells within the brain, but by special sense organs located at local outposts elsewhere in the circulation. The sensitive collaboration of these two systems makes a further contribution towards achieving favourable monotony of the internal environment.

The cells which control these adjustments activate them whenever their immediate environment varies detectably, but they cannot identify the cause of the variation. Like a thermostat which automatically switches on the furnace whenever its thermometer drops below a certain point, whether the cause is the onset of winter or a carelessly opened window, they act on the assumption that, regardless of the events responsible, the situation will be corrected by changing the rate and depth of ventilation.

The commonest cause of increased respiratory exertion is the exorbitant consumption of oxygen which results from strenuous muscular work. Prolonged muscular activity creates an enormous demand for energy: this can be provided

The barrel chest of long-standing chronic bronchitis. Chronic infection destroys the respiratory surface of the lung. To compensate for the loss, the chest ventilates more strenuously and after many years this leaves its mark on the shape of the thorax.

only by the efficient metabolic degradation or burning of carbohydrate fuel; and, since oxygen plays an indispensable part in this degradation, the blood flowing through the working muscles soon becomes depleted. When the depleted blood reaches the respiratory centre in the brain, it automatically sets in motion the extra breathing which is needed to overcome the metabolic debt. As the debt is paid back, the respiratory effort slowly subsides. Under normal circum-

The basic flow diagram of the human heart and lungs.

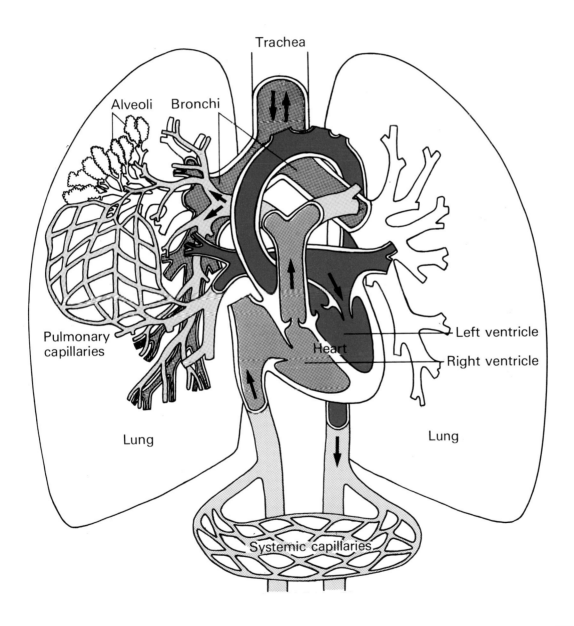

Trachea

Alveoli Bronchi

Pulmonary capillaries

Heart

Left ventricle

Right ventricle

Lung

Lung

Systemic capillaries

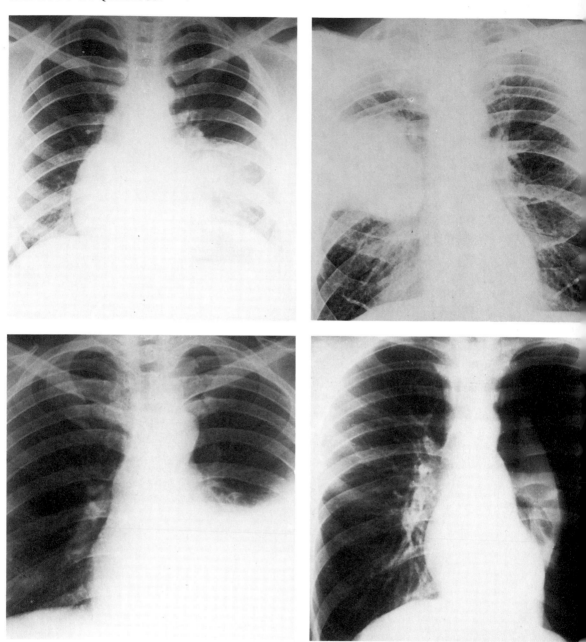

stances, breathing imperceptibly keeps pace with our exertions, and the extra effort is more noticeable to others than it is
to us. There comes a point, though, when the panting becomes intolerable; at that moment we are brought to a standstill by breathlessness.

Some typical X-ray appearances of lung disease.

Top left   *The left lower lobe of the lung has become consolidated as a result of pneumonia. This area would also be dull to percussion but the voice and breath sounds are characteristically exaggerated.*

Top right   *A large solid cancer of the right middle lobe of the lung.*

Bottom left   *A pleural effusion. Inflammatory fluid lying between the lung and the chest wall on the left side When the doctor taps over this area it produces a characteristically dull sound. The sounds of breath and voice are diminished over the same area.*

Bottom right   *Pneumothorax. Air has entered the pleural space and the left lung has collapsed as a result. The blood vessels of the fully extended right lung can be seen reaching to the very edges of the thorax.*

This means that breathlessness is the uncomfortable awareness of strenuous respiration. But the mere fact that it is uncomfortable does not mean that it is a medical symptom like pain, dizziness or nausea. The breathlessness which accompanies strenuous exertion is proof that the body is efficiently recognising and satisfying its own needs. If it occurs without such an obvious explanation — that is to say, in a normal atmosphere and without our making any undue effort — it is reasonable to suspect that the oxygen supply is being frustrated at some point between its arrival in the lungs and its consumption by the tissues, and since the delivery of oxygen and the elimination of carbon dioxide require the close collaboration of heart, lungs and blood, the clinical diagnosis of breathlessness calls for the close investigation of all three systems. What is wrong becomes intelligible only in the light of what is known to be right. The subjective experience of abnormal function can be treated only when one has identified the aims and purposes of normal function.

5 · The Pump

HUMAN BEINGS HAVE ALWAYS ACKNOWLEDGED THAT there is an association between existence and respiration. The start of life is traditionally identified with the first breath, and until more reliable signs of death were recognised, the misting of a mirror or the stirring of a feather were accepted as conclusive evidence that life still lingered. Most death-bed scenes include detailed accounts of the penultimate changes in respiration: the mourners seem to hang on the long breathless silence, catching their own breath until the dying patient sighs once more.

On the previous pages *The Anatomy Theatre at Padua.*

Judging by the number of ways it is mentioned in common speech, it would seem that the heart plays a comparable part in our conscious experience of human life. But although this organ is said to swell, leap, sink, ache and break, we are aware of its existence only because we have learnt to use such phrases. Admittedly, there are many other feelings which can be shown to arise from the heart — flutters in the chest, pounding in the neck, rhythmic roaring in the ears, throbs, syncopations and unsettling percussions. But these bodily sensations which accompany dread, rage or passion do not convey the impression of a single, localised organ whose disordered action is responsible for such feelings. Once we have been entrusted with this knowledge, however, we tend to report such sensations as if their origin were self-evident.

It is almost impossible to think back to the time when the heart had no place in the collective imagination, when clinical wisdom and anatomical knowledge had not yet organised these feelings into a picture. But there must have been a time when the various diseases which were later shown to be associated with the heart were experienced as individual sensations — a flutter here, a throbbing there, breathlessness perhaps. What reason could anyone have for thinking of these orphan twinges as anything other than what they were? How did the heart make its début? Presumably when someone first opened the chest and found it there. That was the least that had to happen. But that alone would not have been enough. Taking a look doesn't automatically reveal what there is to be seen: the innards are not labelled and arranged on shelves, and because we have inherited a clear-cut inventory of the parts we own it is easy to forget that there was a time when no one knew that there were such parts, and certainly did not know one from another — 'this one' as opposed to 'that one'.

Nowadays, a demonstrator can open the thorax expecting to find this and that and, by pointing with his finger, can show what he means. But the visible *thisness* of the heart is not quite so clear-cut as one might imagine. It certainly has distinguishable contours, and abrupt changes of colour — and there would be an extra incentive for regarding that patch of the body as a separate thing if it happened to be independently movable as well. The thing we now call 'the heart' is distinctly beefier than those pink spongey things which seem to fall back from it on either side when the chest is opened. That, at least, is grounds for labelling it 'the heart', and those 'the lungs'. On the other hand, the heart and lungs are tethered to one another by tubes and membranes, and unless you already had a theory which insisted that there was a significant difference between the object you wanted to pluck out of the chest and all those pink tubes which prevented you from doing so, there would be no particular reason for labelling this 'the heart' and those things its vessels.

Anatomical textbooks give the misleading impression that everything in the chest is immediately distinguishable. In the illustrations, the heart is artificially distinguished from its vessels by a bold graphic outline and sometimes a special colour. The aorta is printed in scarlet; the great veins in sky-blue; the nerves are usually represented in green or yellow. The unsuspecting student plunges into the laboratory carcase expecting to find these neat arrangements repeated in nature, and the blurred confusion which he actually meets often produces a sense of despair. The heart is not nearly so clearly distinguished from its vessels as the textbook implies, and at first sight the vessels are practically indistinguishable from one another. A practised eye can readily recognise the gristly pallor of an artery as opposed to the purple flabbiness of a vein, but what *makes* the eye practised are the theories or presuppositions which direct its gaze — and one of the leading theories of anyone now looking into the chest is the one which says that arteries and veins are different because the blood flows through them in different directions. The colour-codes which decorate students' textbooks are not simply vivid illustrations of what there is to be seen, but graphic conventions which illustrate theories about the function of what there is

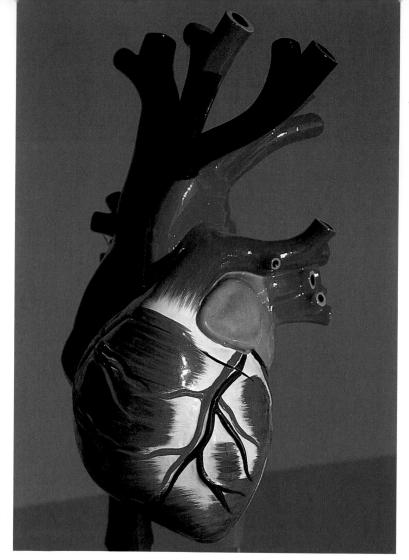

to be seen. And this, of course, was lacking when men first looked into the human chest.

It is tempting to assume that this visual confusion would have resolved itself as soon as passive inspection gave way to active dissection: that the arrangements would have made themselves clear as soon as anyone began to tease apart and display the matted structures. After all, a skilled dissector can separate and free the various vessels which issue from the base of the heart simply by snipping away the sleeves of connective tissues which bind them all together. But what do we mean by a skilful dissector? If we mean someone who can reveal what there is to be seen, that presupposes that he knows what is there to be revealed: his actions are led by his expectations. What makes him skilful is not the steadiness of his hand, but

Student's dissection of a rabbit.

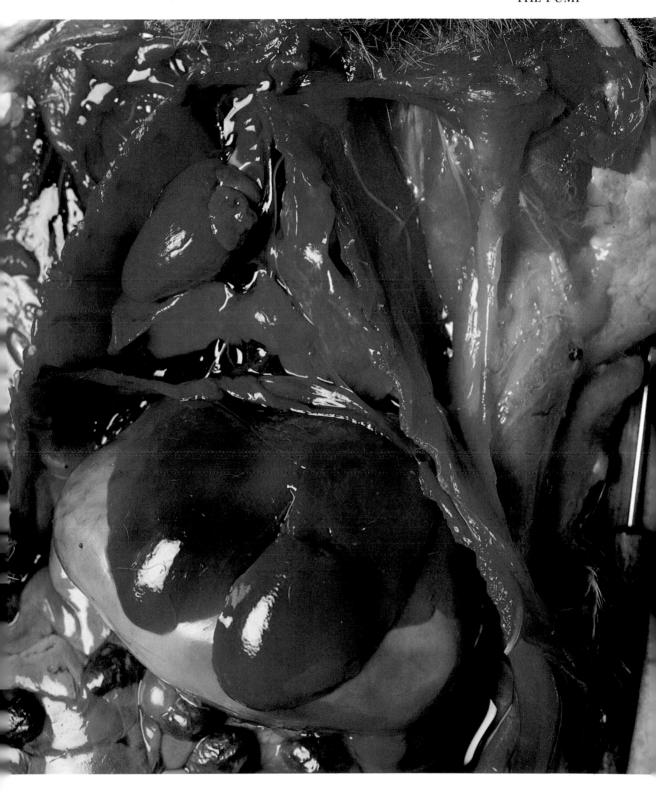

the knowledge which guides it. Knowing where to begin implies some previous idea of how one wants to end up. Otherwise, why slice here rather than there?

The difficulties start even before the dissection begins. How would you set about opening the chest to get the best view? What would count as the best view? By now there is a canonical viewpoint which displays all the known structures to their best advantage and does the least damage on the way in. Any incision is likely to injure some underlying structure, but since the modern investigator knows what is at risk, he can choose the safest incision and, as he already knows which structures he will be sacrificing by going in here rather than there, he can make allowances for what he has destroyed *en route*.

Once inside, the problem becomes even more complicated. Snipping away a sleeve of connective tissue implies that you have already recognised the difference between a trivial covering and the essential structures which lie underneath. Looking at it for the first time, though, you might easily regard a bundle of vessels as a single structure, and the act of unbinding them might seem destructive rather than demonstrative. You cannot know how deeply to cut and at what angle to hold the scissors until you have some knowledge of what you want to show and what there is to damage. The very choice of instruments is an implicit acknowledgment of what you are after and of the risks you are running. When someone favours a blunt probe it is because he knows the structures which would be endangered by the use of a sharp scalpel. There is no such thing as 'taking care' in the abstract: care and caution are defined by the known risks one is trying to avoid, by knowing what would count as having gone wrong. How could a primitive investigator have known this? Even with textbooks to guide you, it is not always easy to recognise when you have made a mistake. An accidentally cut vessel may gape for a moment, giving some indication that one has made a blunder, but if it is sliced clean across, the two ends may retreat into the ooze and mess of the surrounding tissues, and, unless there were good reasons for pursuing them, these elastic fugitives might get left out of the final count.

Considering all these difficulties, it is amazing that any progress was made, although, given time, persistence and conscientious curiosity, the regularity of the layout would have eventually made itself clear. But that begs another question. Since the first appearances are so discouragingly confusing, why would anyone want to persist? It is only to us that the answer seems obvious: to find out how it all works. But the very idea that 'it' — all that stuff inside — was something that worked or performed is a theoretical assumption which would not necessarily have occurred to the naïve observer. It is easy to understand a vague curiosity about one's own interior, a wish to take stock of all those red slippery things which are sometimes revealed as a result of accidents and wounds. But why should anyone think of them as 'the works' as opposed to 'the contents' — as a system rather than an inventory? Why should the notion of 'the works' apply to man at all? Our actions, our perceptions and our liveliness seem self-explanatory. We can do most of the things we want to do, and we can feel, see, hear, smell and touch all that we need to. We are in business. We are in charge. Events are in hand. Things get done. For the normal unself-conscious person, for the person who is satisfied to get on with it, the idea that there might be something which mediated or mechanised his getting on with it would seem bizarre.

The moment when man first suspected that what he did was the result of hidden things getting done must have changed his whole view of the sort of thing he was. The suspicion that his effectiveness or agency was caused by something other than his conscious urge to be effective, and that he himself had no real control over this, constitutes an almost inconceivable leap of the imagination, and one can only conclude that it was largely the result of drawing an analogy between himself and his own technological artefacts. In primitive societies, where technical images are few and far between and very simple at that, most explanatory metaphors are drawn from nature. In the effort to understand his own make-up, primitive man inevitably resorts to images of wind and water, breezes and tides, floods, fruits and harvests. But the development of technology created a new stock of metaphors — not simply extra metaphors, but ones altogether different in their logical character. Once man succeeded in making equipment which performed — looms, furnaces, forges, kilns, bellows, whistles and irrigation ditches — he was confronted by mechanisms

whose success or failure depended on the efficiency of their working parts: things which could block or break, silt up or go out, mechanisms which were intelligibly systematic and systematically intelligible. By mechanising his practical world, man inadvertently paved the way to the mechanisation of his theoretical world.

The success of modern biology is not altogether due to the technology with which we pursue it; the number of technical images we now have for thinking about it play an almost equally important part. An American scientist once said that the steam engine had given more to science than science had to the steam engine, and the same applies to telephone exchanges, automatic gun-turrets, ballistic missiles and computers. Whatever these devices were designed to do, they have incidentally provided conjectural models for explaining the functions of the human body. And the effectiveness of this process grows by geometrical progression. After all butchers,

Human beings have made machines which help them to handle the world. The principles by which these devices work also provide the metaphors with which human beings begin to understand

themselves. It would be foolish to suppose that man is a machine, but by comparing himself with them he gains a useful insight into certain aspects of his own nature.

priests and augurers had been disembowelling animals since the dawn of time, and the battlefields of antiquity would have given ample opportunity for looking inside the human body. One of the reasons why the anatomy and physiology of the heart took so long to develop was the lack of satisfactory metaphors for thinking about what was seen. Of course, the written evidence of early Greek science is so fragmentary that it gives an unreliable picture of what was known or believed. Modern scholars continue to disagree about the authorship of these tantalising fragments, and since much of it takes the form of poetry rather than discursive prose it is almost impossible to get a coherent picture of what the sixth-century BC Greeks knew. The organs contained in the chest had already been identified, and the idea that blood ebbed and flowed in the various vessels was also appreciated, but there was no consistent doctrine and no argued conviction about its mechanism. Until the end of the fourth century BC, no one distinguished between arteries and veins, and even then the distinction didn't carry the systematic significance that it does now.

It would be a mistake to assume that early Greek physiology was as incoherent as the ruined evidence might lead one to believe. Galen inherited a vast treasury of texts and, although most of these are lost to us, the way in which he repeatedly acknowledges the work of his ancient predecessors implies that he was not the first scientist to visualise the blood vessels as part of an intelligible working system. Nevertheless, the weight and cogency of what he had to say is so immeasurably greater than what came before that one is forced to conclude that Galen had some peculiar advantage over his Greek predecessors. And although the unprecedented naturalism of the Roman art of his time indicates that an intense interest in the appearance of the physical world had an important part to play, it is not unreasonable to deduce that this advantage was connected with the technological ingenuity and richness of Roman civilisation.

The exact date of Galen's birth is unknown — perhaps AD 130 — in the Greek city of Pergamum. He received his education in philosophy and medicine first in Smyrna, then in Corinth and finally in Alexandria, where anatomy and physiology had flourished 300 years before. He returned to Pergamum, and became a surgeon to the gladiators — which would have given him a good opportunity to see gaping wounds and flowing blood. He came to Rome in AD 169 and was appointed physician to the pagan emperor Marcus Aurelius. His literary output was enormous: although many of his manuscripts were destroyed by fire in the year 192, his surviving texts amount to more than 9,000 pages devoted to philosophy, medicine and physiology. Much of the work consists of careful commentaries on the work of his great predecessors. He believed that all progress took its origin from ancient wisdom, and although he was committed to personal observation and experiment — he insisted on skinning his own animals rather than leaving this menial task to a slave — he always conducted his enquiries in the context of the great tradition which had been established by Hippocrates, Plato and Aristotle, and by the Alexandrian physicians of the fourth century BC. He inherited and consolidated the traditional

During the Christian Middle Ages the scientific work of Galen was largely ignored in Western Europe but scientific medicine was unable to advance until his theories were revived, criticised and finally superseded.

notion that the universe was composed of four basic elements, and that these represented the four irreducible qualities of wetness, dryness, warmth and cold. He agreed with the group of writers who are now included under the name 'Hippocrates' that the four universal elements were reproduced in the human body by a quartet of physiological 'humours', and that these took their origin from the elements to be found in food and drink.

For Galen, in fact, physiology started with nutrition. Nutrition was an act of creative transformation which systematically converted the inanimate elements into the active ingredients of the living body. In the first stage, food was absorbed from the gut and passed via the portal vein into the liver, where it was brewed or concocted into blood and at the same time imbued with a weak form of vital principle which Galen called 'Natural Spirits'. On leaving the liver, this raw material ran in an indolent current through the great veins, which distributed it to the hungry tissues. Depleted by the tissues of its Natural Spirits, the blood sluggishly regurgitated back along the same channel, where it was replenished with fresh nourishment by the liver.

Galen recognised that although this theory explained how the body maintained its bulk, some extra principle was required to account for its warmth, verve and spontaneous reactions. The Natural Spirits concocted in the liver might be enough to explain the vegetative functions of the body, but Galen insisted that the blood which went to the brain had to be supercharged with extra force or energy. This, he suggested, was obtained from the air for, like his predecessors, he regarded the invisible element which filled the atmosphere as a vital force.

And this is where the heart enters the system: as the creative mediator between blood and breath. According to Galen, some of the blood leaving the liver overflowed or was diverted into the right side of the heart, where it came into immediate contact, or so he supposed, with air, which had been drawn in through the lungs and along the pulmonary veins. The encounter produced a mysterious incandescence, a biological flame which heated the blood and gave it the fertile warmth which is so characteristic of the living body.

Galen likened the heart to a lamp, fed by the oily fuel

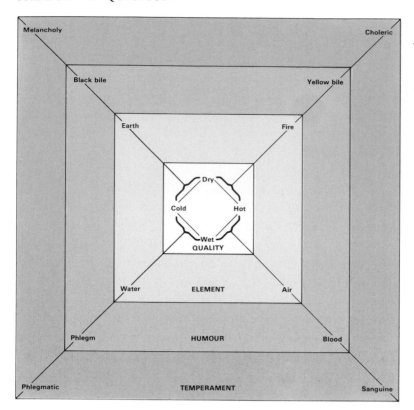

The physiological theory of the four humours was integrated into an overall cosmology which related qualities, elements, planets, seasons, ages and winds.

supplied from the liver. The smoky fumes exited through the flues of the pulmonary veins and the windpipe — anyone could see that breath steamed or smoked in the cold air. But it was more than a heating-system. The blood was not only consumed by the cardiac fire, it was transformed and refined: converted from the thick, purple ooze supplied by the liver into the swift, scarlet stream which issued from the arteries. In this respect, the heart resembled a smelter's furnace, burning off the combustible impurities which were present in the food, leaving a purified residue which was enriched with the peculiar pneumatic principle said to be present in air. The leaping red fountain which shot from the left side of the heart differed from the turbid material entering it from the right not simply in lacking combustible impurities, but in possessing a weightless substance which Galen called 'Vital Spirits'. This superlative substance was distributed through the arterial system, some of it going to the general tissues, where it presumably reinforced the effect of the Natural Spirits, and

186

some of it going to the brain, where it underwent further refinement and became Animal Spirits — the material responsible for converting thought into action.

This elaborate theory brought together and reconciled the scattered theories of antiquity, organising them into an intelligible system: an industrial plant half-way between a brewery and a blast-furnace. It seemed to make sense not only of the various pipes and tubes which were known to exist, but of the various material intakes and outputs of the living body. It explained the source, function and distribution of nourishment, the purpose of breathing, and the spontaneity of the living body. Its dependence on technological metaphors is self-evident: the most notable feature of the system is the emphasis on manufacture and transformation, cooking, brewing and smelting — processes which convert, purify and refine tangible substances. The heart, like the liver, is simply another part of the factory.

The recognition of the propulsive role of the heart was delayed for nearly 1,500 years, although the necessary evidence was just as available to Galen as it was to William Harvey. The difference between the two men is not one of ingenuity and skill — in fact, if these were the sufficient conditions of scientific progress, Galen rather than Harvey might have been the discoverer of the circulation of the blood. But seeing is not all that there is to believing; belief determines the significance of what is seen. The difference between Harvey and Galen was one of metaphorical equipment. When Galen tried to systematise the relationship between the blood and the breath, the co-operation between the liver, the lungs and the heart, the processes on which he modelled his theory were the most conspicuous features of the world in which he lived. There were no better analogies than those of the lamp or the smelter's furnace: they were the most intelligible images of transformation and change. One can only assume that Galen's inability to see the heart as a pump was due to the fact that such machines did not become a significant part of the cultural scene until long after his death. The heart could be seen as a pump only when such engines began to be widely exploited in sixteenth-century mining, fire-fighting and civil engineering.

In the absence of a more plausible metaphor, Galen's

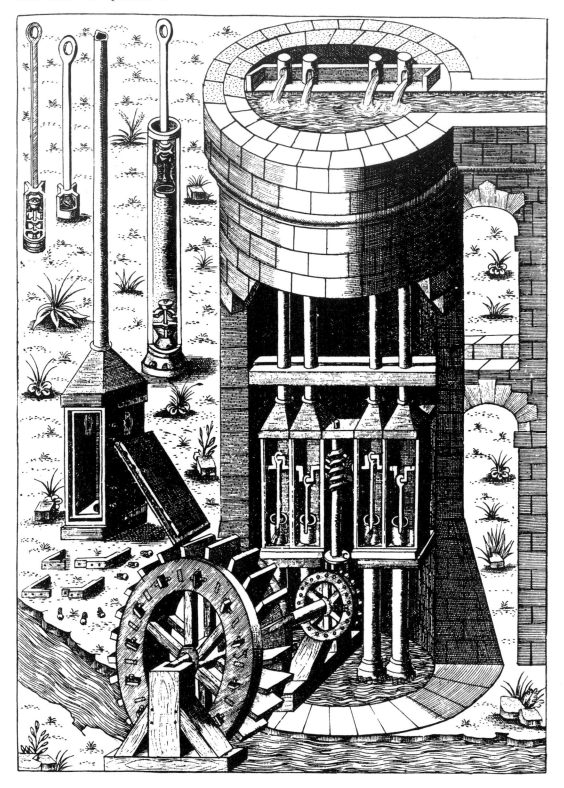

The growth of mining during the sixteenth century led to a great increase in the use of pumps. Knowledge of these devices may have played an important part in seeding Harvey's fertile imagination.

industrial model inevitably monopolised the imagination of late antiquity, blinding men to inconsistencies that later became self-evident. For instance, the furnace model required the blood to move directly across the heart from right to left, in spite of the fact that its passage is quite obviously blocked by the thick muscular wall which separates the two ventricles. We now know that the only way for the blood to get from one ventricle to the next is by going the long way round through the lungs, and that, as we saw in the previous chapter, the transformation from purple to scarlet takes place there rather than in the heart. Galen's system gave little or no emphasis to the pulmonary circulation, dismissing it as a lubricant trickle which nourished the bellows of the cardiac furnace. Since his theory depended on a direct transit across the heart, he insisted that the septum which divided the two ventricles was perforated by channels. He admitted that these channels were too small to allow the passage of thick, unrefined blood, but since his theory also insisted that the blood was refined by its encounter with air this did not pose a problem.

This casuistry shows how persuasive a metaphor can be, and how, once an idea lodges in the imagination, it can successfully eliminate or discredit any evidence which might be regarded as contradictory. No special technique was needed to show that the septum is impermeable: although its surface has an irregular, corrugated appearance which might give the impression that it is pitted with the mouths of small vessels, a quick poke with a bristle would have shown that these pits led nowhere — certainly not from one side to the other. Galen could have performed this experiment just as easily as his successors did. The great sixteenth-century anatomist Vesalius did perform it, but since he, too, was under the spell of the traditional theory, he insisted that the blood 'sweated' across the muscular wall through pores which were admittedly too small to allow the passage of a bristle.

Philosophers of science sometimes imply that scientific thought is a simple alternation between conjecture and refutation, and that contradictory evidence automatically discredits an otherwise plausible hypothesis. The history of cardiac physiology shows that this is an over-simplification, because it overlooks the criteria which are used to decide what will count as a contradictory finding: if a theory has found favour with a scientific community, it is the anomalous finding rather than the theory itself which is discredited. This is exactly what happened in the experiment with the bristle. It was only when the pulmonary circulation was accepted as an established fact that the results of this experiment could be recognised as significant. Meanwhile, as is so often the case, it was more convenient to make an *ad hoc* modification of the existing theory. The history of science is full of such examples. If a theory is persuasive enough — and the reasons for any given theory being so persuasive are often hard to list — scientists will accommodate inconsistent or anomalous findings by decorating the accepted theory with hastily improvised modifications. Even when the theory has become an intellectual slum, perilously propped and patched, the community will not abandon the condemned premises until alternative accommodation has been developed.

The same principle applies to the problem of blood flow in the veins, especially to the valves which are to be found in these vessels. Galen could have seen these structures just as easily as his successors did. You have only to open the veins to see that there are little flaps on the inside wall and, from the way these flaps are facing, their valvular function now seems obvious. To us, as to Harvey, they are the most significant contradiction to Galen's theory that the blood could flow hither and thither in the veins: as Harvey saw, the flaps are arranged so that the blood can flow only one way through them. He proved that the flow was one-way by a dazzlingly simple experiment. He tied a tourniquet round the upper part of his arm, just tight enough to prevent the blood flowing back to the heart through the veins but not tight enough to prevent blood entering the arm through the arteries. The veins swelled up below the tourniquet and remained empty above it, which implied that the blood could be entering them only through the arteries. By carefully stroking the blood out of a short length of vein, he saw that the vessel filled up only when the blood was allowed to enter it at the end which was furthest away from the heart.

These experiments are so simple that it seems surprising that they weren't performed before. But it would not have occurred to anyone to perform an experiment like this unless

Harvey's famous experiment with the veins in the arm seems deceptively simple. But simple experiments are often dictated by complex theories. Without such a theory no one would have been interested in performing an experiment like this.

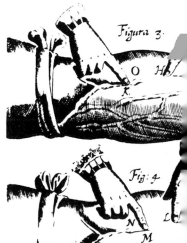

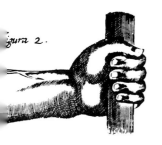

he suspected that there was something wrong with the traditional theory, or that there was an alternative theory whose implications required such an experiment as confirmation. But alternative theories can be conceived only against a background of well-established tradition. One of the reasons it took so long to overthrow the Galenic theory was not that men were overawed or enslaved by it, but that they were not fully acquainted with it. For more than 1,000 years after Galen the bulk of Greek scientific literature was either lost, forgotten or neglected, and most of Galen's work remained untranslated until the revival of ancient learning marked the beginning of the Renaissance. A theory which survives uninterruptedly has quite different consequences from one which is revived after a long period of neglect (this also applies to art or literature — imagine how different our attitude to Shakespeare would be if his plays had been taken out of commission between 1700 and 1900). When theories are vigorously discussed, examined and contradicted for any length of time, they tend to evolve of their own accord. In fact, as long as a theory continues to occupy the public domain, it automatically sets up reactions which lead to its own development and eventual downfall: its uninterrupted existence, its acknowledged claim to be a subject of controversy, guarantees not so much its survival, as its inevitable replacement by more convincing alternatives.

Galenic physiology was never given the opportunity to undergo the ordeal of unremitting criticism, so that when it was re-examined in the sixteenth century it had not yet spent its powers of persuasion. Because of the ascendancy of Christianity, for more than 1,000 years interest in the nature of the physical world was supplanted by a passionate concern with the metaphysical fate of mankind. The physical order of nature seemed trivial by comparison with what its members had been offered by the redemptive sacrifice of Jesus Christ. The biological origins of the human individual paled into insignificance in the light of what was now known about his theological destiny.

If one compares Roman art of the pagan second century AD with the European art of the next 1,000 years, it is easy to see how much Christianity changed men's attitude to the world in which they lived. The character of the Pompeian murals, for instance, seems quite consistent with Aristotelean natural history. They show a world where the appearance of fruit, flowers and animals was delightful, and man was obviously pleased to linger: congenial people sit in natural attitudes — windows give on to scenes where trees shimmer in the wind, and distance fades convincingly — bodies and objects are solid, and the volumes they occupy, hospitable and well-proportioned — even the atmosphere has physical existence. A civilisation which took so much pleasure in how the world looked was bound to become curious about the way it worked — it is no accident that these pictures, with their spectacular and seductive realism, are contemporary with the encyclopaedic biology of Galen.

Less than 500 years later, the scene is changed beyond recognition. The human figure is now reduced to a flattened silhouette, almost always draped or clothed — even when the naked body of the crucified Saviour is shown, his physical features are little more than hieroglyphs. Bodies are pinched and bloodless; they have no surface, no skin, blood, or muscles. It is not that they are insubstantial in the way that phantoms are — it is as if substance were an altogether irrelevant consideration. Faces are diagrammatic and generalised, and the only way of distinguishing one saint from another is by the emblems or attributes which float like ornamental kites in the vicinity of each. The evangelists perch awkwardly on their cramped thrones, hands twisted towards improbable inkwells; their clothes curdle rather than drape about their notional limbs; the rooms or alcoves in which they preside are spartan sentry-boxes, unfurnished, airless and unlit; the backgrounds are opaque, gilded or diapered with abstract ornament; there is no impression of a world beyond, and on the rare occasions when topographical features are allowed to appear, they invariably take the form of theatrical props — trees, plants and animals have dwindled into decorative ciphers. These pictures are not simply incompetent versions of this world — they are devotional aids for a society impatient for the next one.

At the end of the fourteenth century, the visual arts begin to breathe with life again: air and light are let in; pleasing vistas reappear; limbs plump out, fatten, flex and fill with blood. The spectator's glance is now allowed to travel onwards into the views beyond: there is vapour and blueness, and in the

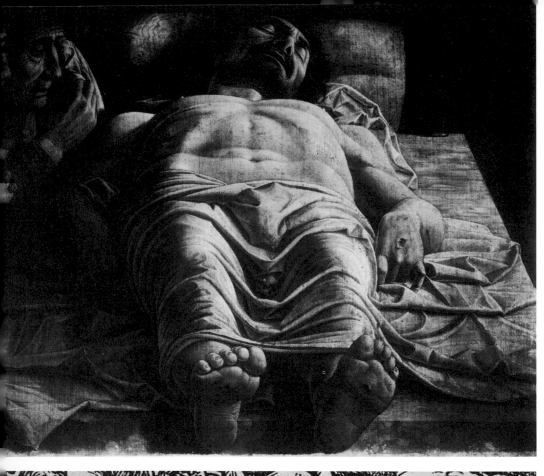

By the fifteenth century curiosity about physical appearances had become so intense that painters took pains, if not pleasure, in representing the surgical details of Christ's wounds. (Master of Flémalle, Städelsches Kunstinstitut, Frankfurt-am-Main)

Opposite, above *Mantegna's dead Christ. (Brera, Milan)*
Opposite, below *Detail from the frontispiece of Vesalius's great textbook on human anatomy.*
    During the early Christian era interest in the Natural Man gave way to a concern for the fate of the new creature created by baptism. The details of the physical body seemed less important than the generalised scheme of the spiritual person. From the thirteenth century there was a slow revival of interest in the appearance of mortal things, and by the fifteenth century an artist such as Mantegna is able to represent the Saviour as a particular corpse.

miniatures of Jean Pucelle and the Limbourg brothers men move freely once again in a seasonal world. With newly discovered glazes, painters take pleasure in showing the play of light on fur, flesh, metal and jewellery: pearls gleam, and diamonds flash; artists pride themselves on their ability to represent the different textures of ermine and velvet.

When hand and eye are led into the world like this, the mind is soon drawn in as well: curiosity revives and science can resume its interrupted task. But curiosity alone is not enough; proper science needs an aim; a particular interest; a problem. After a long period of intellectual hibernation, the only available problems were hidden in long-forgotten manuscripts.

195

It is one of the paradoxes of scientific history that the great leap forward took place only when scholars went back to the scientific texts of antiquity. The same principle applies to the history of art. Degas, for example, is regarded as one of the great innovators, but he was an almost obsessional copyist of

Ingres, 'The Martyrdom of St Symphorian', 1834. (Fogg Art Museum, Harvard University)

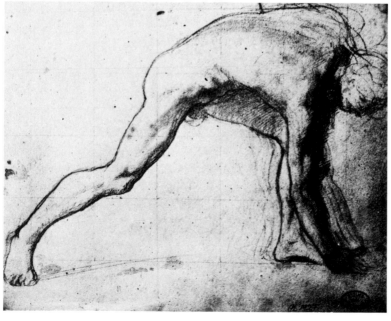

Degas, 'The Daughter of Jephthah', 1859.

198

classical drawings, perhaps the most skilled copyist there has ever been. He insisted:

> Museums are there to teach the history of art and something else as well. They stimulate a desire in the weak to imitate but they furnish the strong with the means of their own emancipation. The painters of the 16th century are the only true guide. If you are thoroughly acquainted with them and if you ceaselessly perfect your means of expression with an unflinching study of nature, you are bound to achieve something.

It is this tension between tradition and fresh observation which marks the work of the Belgian anatomist Andreas Vesalius. In 1543, the year that Copernicus issued his work on the revolution of the heavenly bodies, Vesalius published the first modern textbook of anatomy. The illustrations are so beautiful — they were probably done by a student of Titian — that it is easy to overlook the text, and yet it is here that the connection with Galen is recognisable. Medieval textbooks of anatomy have no systematic scheme, whereas Vesalius closely followed Galen's method and reconstructed the human physique from the foundations of its skeleton: 'As poles are to tents and walls are to houses so are bones to living creatures.'

But Vesalius also set himself the task of surpassing the work of his great predecessor and, as Degas later advised, he undertook to perfect his means of expression by an unflinching study of nature. But although he had the advantage of working with human corpses, where Galen had to make do with barbary apes, in his interpretation of the function of the heart he reproduced all the errors of the ancient wisdom.

In the first edition of his book he repeated Galen's claim that air entered the left ventricle of the heart, where it served the double purpose of cooling the innate heat and preparing vital spirits. Like Galen, he insisted that the blood soaked plentifully through the inter-ventricular septum, and that, although this wall was one of the thickest parts of the heart, it was perforated throughout by little channels. By the time Vesalius was ready to publish a second edition, he had begun to entertain serious doubts about the existence of such perforations. These doubts may have been reinforced by the emergence of a theory which made such a direct transit

Titian's portrait of the Belgian anatomist Vesalius. (Galleria Palatina, Florence)

Shortly before William Harvey arrived in Padua, Fabricius of Aquapendente, the Professor of Anatomy, designed and built a new arena in which to demonstrate human dissection. This is not a room so much as an optical device, focusing attention on an object of common interest. In arenas shaped almost exactly like this the human body had once been mutilated for the entertainment of Roman citizens.

unnecessary. Shortly before Vesalius published his second edition, the Spanish philosopher-theologian Michael Servetus had suggested that the blood made its way from right to left through a north-west passage in the lungs. This physiological theory appears in a Unitarian treatise for which Servetus was burnt at the stake in Geneva by Calvin in 1556.

Given the traditional relationship between air and the soul, it is not altogether surprising that a theologian should have concerned himself with the physiology of breathing. Like his predecessors, Servetus recognised that inanimate matter had to be galvanised by some special principle, but he believed that the fertilising encounter between air and blood took place in the lungs, not in the heart. Vesalius probably read the heretical treatise, and that may have been what prompted him to perform his experiment with the bristle. It was left to one of

William Harvey. The first modern biological scientist. After a preliminary medical course at Cambridge Harvey graduated from the University of Padua, where he was taught by the great anatomist Fabricius. Fabricius's demonstration of valvular flaps in the veins later played an essential part in Harvey's theory of the circulation of blood.

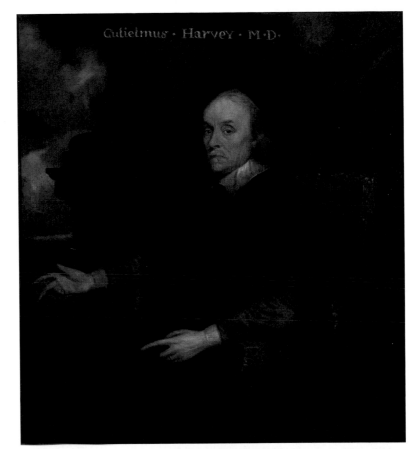

Harvey's medical diploma from the University of Padua.

his students to make the conclusive break with the traditional theory.

In 1559, Realdus Columbus published a book entitled *De Re Anatomica*. In terms of pure anatomy he had very little to add to the work of his great teacher, but, when it came to understanding the function of the lungs, Columbus was so imaginative that William Harvey some ten years later copied out the following passage from his book:

Everyone thinks that there is a way open for the blood to pass from the right ventricle to the left, and that this may be more easily accomplished, they think that it is refined in the transit … But they err by a long way, for the blood is carried to the lung through the pulmonary artery and in

Illustrations from Vesalius. The 'De Fabrica' was published in the same year as Copernicus's book on the revolution of the heavenly bodies. If Copernicus had displaced man from the centre of one universe, Vesalius had put him right at the centre of another one.

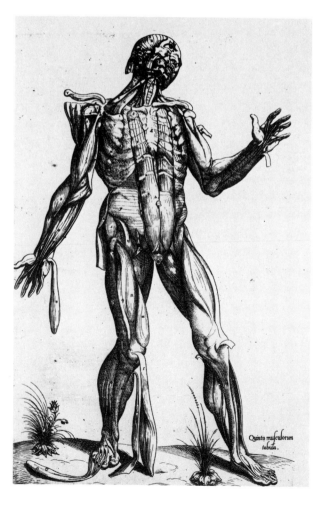

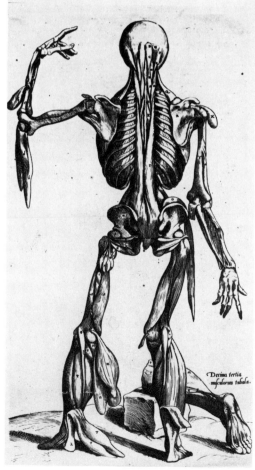

the lung it is refined, and then together with the air it is brought through the pulmonary vein to the left ventricle of the heart.

Columbus pointed out that Galen's theory not only required blood to pass across an impermeable septum, but also implied the presence of air and smoky fumes in the pulmonary blood vessels. He showed that the pulmonary vessels were full of blood and there was no way for air to pass from the lungs into the heart: the only way was for the blood to visit the air in the lungs.

The publication of these ideas did not automatically lead to the downfall of Galen's theory. There are certain moments in the history of science when the introduction of a valuable truth merely adds to the confusion. At the end of the sixteenth century, physiology was in the same condition that Copernicus had found astronomy in nearly sixty years before: 'It is as if an artist were to gather together the hands, feet, head and other members from many different models, each part excellently drawn, but not related to a single body. And since they in no way match one another, the result is more like a monster than a man.'

Although Harvey was aware of Columbus's dissenting opinion, he was unable to take advantage of it until he had reconsidered some of the other traditional assumptions about the actions of the heart. For example, the official theory claimed that the heart filled itself like a syringe, by actively expanding and drawing the blood up from its natural source in the liver to a level where it spontaneously overflowed into the veins. As the overflow escaped, the heart shrank until it was ready to re-expand, creating a vacuum which would suck in a fresh supply of blood.

This theory takes little or no account of the heart's muscular contraction, and although it was generally accepted that the heart-beat has two phases — swelling followed by shrinkage — the traditionalists thought that the heart was only truly active at the moment when it was expanding, and that the contraction was simply a question of elastic recoil. To some extent, this misunderstanding came from the fact that classical physiologists knew next to nothing about the mechanics of muscular contraction, or even that there was such a thing. If you watch your own biceps contracting, it is easy to get the impression that it is increasing in volume. The fact that such a mistake can be made about a muscle which can be moved slowly at will explains why it was so difficult to identify what was happening in an organ which was beating seventy times a minute. When Harvey first opened the chest of a living mammal, the heart-beat was so rapid that he was unable to tell the difference between expansion and contraction, diastole and systole, let alone identify the active phase:

> I kept finding the matter so truly hard so beset with difficulties that I all but thought ... that the heart's movement had been understood by God alone. For I could not rightly distinguish ... [when] or where constriction and dilation occurred. This was because of the rapidity of the movement, which in many animals remained visible for but the wink of an eye or the length of a lightning flash, so that I thought I was seeing now systole from this side and diastole from that side; now the opposite; the movements now diverse, and now

Cardiac muscle. Unlike muscles which move limbs, heart muscle is spontaneously active and fibres like these will continue to beat in isolation. Harvey would have been mystified by this and attributed the heart's action to some vital principle which was present in the blood.

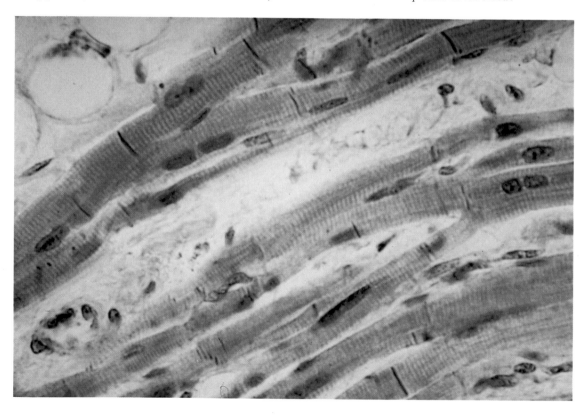

inextricably mixed. Hence my mind was all at sea and I could neither come to a decision myself nor assign definite credit to others.

Recognising that the problem was one of speed, Harvey imaginatively switched his attention to animals whose heart-beat was slow enough to let him see the two phases one after the other: cold-blooded creatures such as toads, serpents, frogs, snails, lobsters and prawns. He also studied the dying hearts of warmer animals, finding that as they began to flag and move more languidly he could inspect and determine the nature of the movement.

From these observations, he concluded that the heart was active not when it was expanding, but when it was most vigorously shrinking, when its walls were thickening. This, he decided, was the moment when the heart-beat made itself visible on the outer wall of the chest. By cutting the arteries leading out of the heart, he discovered that the blood was ejected into them at the moment when the walls thickened and bunched up, and that the blood re-entered through the veins as this active contraction relaxed. He pointed out that the active phase was one of contractile movement and that, contrary to traditional belief, the heart filled itself following the moment of its passive relaxation or collapse. These observations immediately made sense of the arterial pulse, for Harvey recognised that the percussion which could be felt at the wrist coincided with the propulsive thrust which could be detected by feeling the outer wall of the chest. Harvey compared the arterial pulse to the effects of 'blowing into a glove and producing simultaneous increase in volume of all its fingers'.

By working with slow hearts, Harvey was also able to see that not all the parts of the heart contracted simultaneously in a synchronised burst, but that there was an orderly sequence of contractile events, which started in the receptive chambers of the auricles and proceeded toward the propulsive chambers of the ventricles:

Those two movements, one of the auricles and the other of the ventricles, occur successively but so harmoniously and rhythmically that both (appear to) happen together and only one movement can be seen, especially in warmer

205

animals in rapid movement. This is comparable with
what happens in machines in which, with one wheel
moving another, all seem to be moving at once. It also
recalls that mechanical device fitted to firearms in which,
on pressure to a trigger, a flint falls and strikes and
advances the steel, a spark is evoked and falls upon the
powder, the powder is fired and the flame leaps inside and
spreads, and the ball flies out and enters the target; all
these movements, because of their rapidity, seeming to
happen at once as in the wink of an eye. In swallowing too
it is similar. The root of the tongue is raised and the
mouth compressed and the food or drink is driven into
the fauces, the larynx is closed by its muscles and by the
epiglottis, the top of the gullet is raised and opened by its
muscles just as a sack is raised for filling and opened out
for receiving, and the food or drink taken in is pressed
down by the transverse muscles and pulled down by the
longer ones. Nevertheless, all those movements, made by
diverse and opposite organs in harmonious and orderly
fashion, appear, while they are occurring, to effect one
movement and to play one role which we style
'swallowing' …

*Quite apart from its standing
as one of the great scientific
texts, Harvey's dissertation on
the movement of the heart and
blood ranks as a classic in
European literature and can be
read with just as much pleasure
as 'Paradise Lost'.*

By now Harvey was convinced that the heart-beat was an
act of vigorous propulsion, and the implications of this conclusion led to the complete overthrow of the traditional
theory. He recognised, for instance, that the amount of blood
propelled into the arteries at each beat was such that if it were
multiplied by the number of beats in an hour the total volume
would be more blood than the body contains. His measurements of the output were relatively inaccurate, but the fact
that he was prepared to see the problem in quantitative terms
at all marks him as the first modern biologist. A more pedantic
man might have got bogged down in finicky measurements
but, like all great scientific geniuses, Harvey saw the implications of even a rough approximation. He recognised that the
hourly output of the heart was so large that the only way it
could replenish itself was by taking in through its back door all
that it had thrown out through the front: 'In consequence, I
began privately to consider if it had a movement, as it were, in
a circle.'

206

It is hard to say when or how Harvey first had this insight. He had been investigating the subject for at least ten years before he published his great book in 1628. Towards the end of his life, he told Robert Boyle that the idea of circulation first occurred to him as a result of seeing valves in the veins:

> In the only Discourse I had with him ... [he said that] when he took notice that the Valves in the Veins of so many several parts of the Body, were so Placed that they gave free passage to the Blood Towards the Heart, but oppos'd the passage of the Venal Blood the contrary way: he was invited to imagine that so Provident a Cause as Nature had not so Plac'd so many valves without Design: and no Design seem'd more probable, than That, since the Blood could not well, because of the interposing Valves, be Sent by the Veins to the Limbs, it should be sent through the Arteries, and return through the veins, whose valves did not oppose its course that way.

But the way in which a scientist remembers and publishes his arguments is not necessarily the order in which the idea originally occurred to him, and since Harvey did not log his daily thoughts we shall never know for certain how the notion of a continuous circulation first occurred to him. Scientists are notoriously forgetful about the origin of their most interesting conjectures, and although the existence of valves confirmed Harvey's hunch that the blood flowed one way through the veins it seems likely that he had already recognised the existence of a one-way circulation, and that this made him realise that the flaps on the inside of the veins *had* to be valves.

As far as one can tell, his most fruitful insight was his recognition of the propulsive power of the heart, coupled with the experiments which confirmed it. By cutting arteries, Harvey showed that the rhythmic spout of blood invariably issued from the end nearest to the heart and that it coincided with the moment when the heart contracted and whitened. By tying ligatures at strategic points throughout the circulatory system, he showed that the vessels became empty and pulseless beyond the block, swollen and engorged with blood behind it.

As I have already pointed out, Harvey would not have been prompted to perform such experiments unless he had already had a hypothesis which forecast their outcome. How did he

207

first entertain an idea which so systematically contradicted all that had been suggested previously? His personal observations of the heart's actions played an indispensable role, and it was certainly a stroke of genius to think of examining slow hearts. But once again the influence of technological metaphor must not be overlooked. By the end of the sixteenth century mechanical pumps were a significant part of the developing technology of Western Europe. Coal and metal mines were being deepened to supply the needs of growing cities. Engineers were bedevilled by the problems of seepage, and forceful pumps were the only way of keeping the shafts empty. Contemporary handbooks of metallurgy included pages of pumping mechanisms. The hydrostatic principles were also applied to the design of ornamental fountains, and in 1615 Salomon de Caus published *Les Raisons de forces mouvantes*, describing a machine for putting out fires:

> The said pump is easily understood: there are two valves within it, one below to open when the handle is lifted up

It is easy to over-stress the coincidence between a given technical invention and a new scientific theory, but the intermittent jet of a fire pump like this would have provided a new way of visualising the heart's action.

Opposite above *Capillaries in the web of a frog's foot. And,* below, *injected capillaries branching throughout heart muscle. Harvey's theory of the circulation of blood presupposed a direct connection between the arteries and the veins. The newly invented microscope allowed Malpighi to see these small vessels but it is quite possible that he would have overlooked or misinterpreted their appearance if Harvey had not primed his curiosity.*

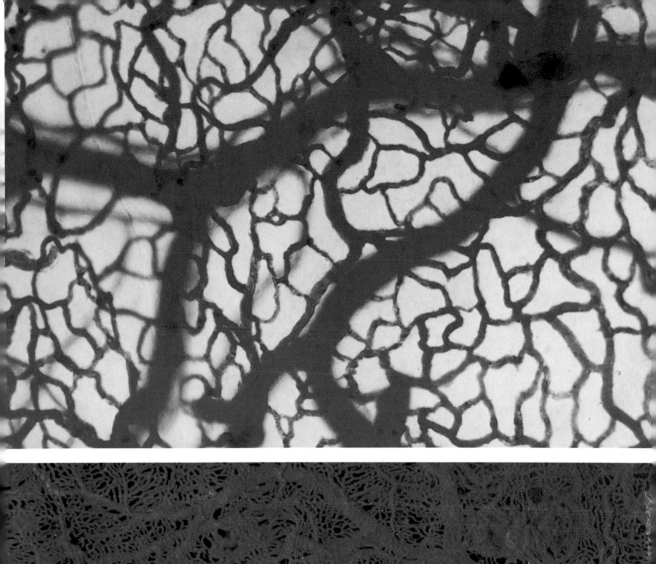
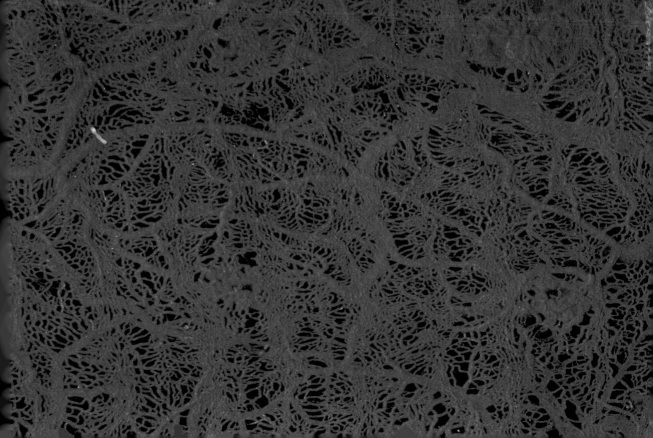

and to shut when it is down, and another to open to let out the water; and at the end of the said machine there is a man who holds the copper pipe, turning it from side to side to the place where the fire shall be.

Historians still disagree about the influence of the fire-pump, but it seems unlikely that Harvey would have departed so radically from the traditional theory if the technological images of propulsion had not encouraged him to think along such lines.

Harvey marshalled all his arguments into a single book of 100 pages. It was originally published in Latin, but even in English translation it has the force of great literature. What strikes the modern reader is the inexorable march of its reasoning, the simplicity of its conjectures and its unflinching determination to develop all its implications in an experimentally checkable form. Naturally, the work is incomplete. It was left to Harvey's immediate successors to find out why the blood circulated through the lungs, but by proving so conclusively that it did he created a soluble problem — which is the least that any scientist can demand of one of his colleagues.

He failed to find the tiny vessels which linked the arteries to the veins, and it was only when the Italian microscopist Malpighi turned his lens on to the lungs forty years later that the existence of the capillaries was even suspected. If anyone before Harvey had discovered these vessels they would have found it impossible to explain their function, and when the function of something is not recognised its visible appearance is often misrepresented as well. Things tend to look like what we know they are for — and if we don't know what they do, we often find it hard to say how they look. Harvey might have discovered these vessels for himself if he had possessed a microscope: his theory demanded their presence, and it is a characteristic feature of fruitful scientific theories that they suggest the existence of objects or processes which may not be discovered until the appropriate instruments reveal their presence.

Like all good theories, Harvey's bristled with unfinished business, and the fact that he was unable to finish the business himself does not diminish his greatness. All subsequent investigations have been framed by his basic assumption, and,

Marcello Malpighi, 1628–94. (A.M. de Tobar, Royal Society)

although Harvey would have been puzzled by the sophisticated electronic and biochemical methods now used to pursue these investigations, he would certainly have understood their significance. He might have been mystified by the anaesthetic techniques which enable a modern surgeon to enter the chest and open the heart, but his own theory dictates the replacement of a damaged valve by an artificial substitute: since he explained the function of the walls which separate the various chambers of the heart, he would have understood and applauded the repair of congenital holes.

It would nevertheless be an exaggeration to claim that Harvey's theory foresaw or even implied *all* that we now know about the heart and circulation. It would be too much to expect any theory to do that. Harvey's hypothesis accommodates but does not actually predict what we now know about the circulation. Harvey recognised that the heart could alter its performance, that it could speed up with exertion and change its rate under the influence of strong emotions, but he did not suspect that such alterations were part of an elaborate homeostatic repertoire, or that the heart and blood vessels, like the lungs, kidneys and every other system in the body, are constantly adjusting their behaviour in order to meet the varying stresses of life. He did not know that the arteries are sleeved with muscular tissue and can, therefore, change their calibre and adjust the volume of the circulation to compensate for blood loss, or that the flow can be redistributed both amongst and within the various organs of the body. It wasn't until the end of the nineteenth century that physiologists discovered the existence of a so-called vaso-motor system, an extensive network of automatic nerves co-ordinated within the brain which regulates the behaviour of the heart and blood vessels.

Harvey's failure to recognise the existence of such mechanisms, and the failure of his theory to predict their existence, are forgivable errors of omission. The function of the voluntary nervous system was not yet understood, so how could he possibly have suspected or even described the functions of involuntary nerves? And since the idea of self-regulation or homeostasis was not to be born for more than 200 years, his failure to appreciate the versatility of the circulation is quite understandable.

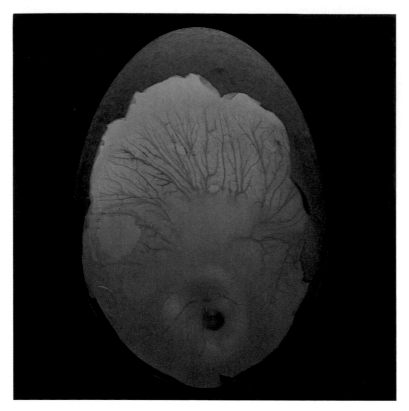

Inside a developing chicken's egg.

The only point where it could be said that Harvey made a positive mistake was in explaining the physiological origin of the heart-beat itself. He claimed that the action of the heart was not spontaneous, but rather arose from the origin of all spontaneity — the living blood. In his investigations of a developing chick, he detected a pulsing spot of blood whose rhythmic flashes preceded the recognisable appearance of the heart. If his lenses had been better, he would have seen that the pulsation was caused not by the blood, but by the transparent heart which enveloped it. His error was not entirely due to faulty observation. Like his colleagues and predecessors, he was susceptible to ancient dogma, and, although he was able to take an altogether mechanical view of the function of the heart, he could not shake off the traditional belief in the existence of vital principles. In the seventeenth century, the spontaneity or 'go' of living things could be explained only by referring back to some spiritual prime mover, and in Harvey's case this energetic agent was not merely to be found in the blood, it was identical with it.

Harvey had no reason to suspect that the mechanical events of the heart-beat were accompanied by electrical changes. New instruments extend the spectrum of visibility. We can now 'look' at the electrical image of the heart in much the same way as modern astronomers can examine the radio image of the stars.

212

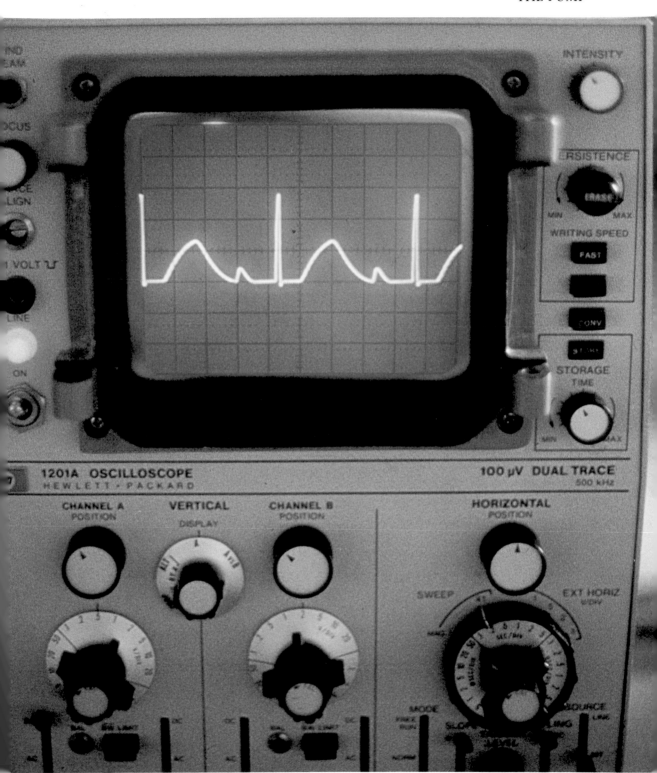

IF BREATH IS THE MOST RECOGNISABLE SIGN OF LIFE, PINKNESS has always been regarded as one of the most reliable indications of health. We instinctively associate redness with vitality, and automatically assume that anyone with a ruddy complexion or red lips is vigorous and robust. The vital importance of blood has always seemed self-evident; from the earliest times man was aware that 'this pure cleare lovely and amiable juyce is the special thing that conserveth every living creature in his being and ... that this treasure of life must most carefully be conserved because it is of all humours the most excellent and wholesome'.

For such reasons blood has always been regarded as a form of natural wealth: a rich liquid asset settled on each individual as a birthright, a priceless deposit which can neither be spent nor accumulated, only lost or dispersed through injury or ill-health. There are no plutocrats, only paupers. Adequacy is abundance.

Medicine took a great leap forward when William Harvey discovered the way in which this precious substance circulated and recirculated through the body of the living individual. Unlike the wealth of a miser, which accumulates without its doing any useful work, the value of blood can be exploited only if it is kept ceaselessly on the move. It is useless when stationary, but it is beyond price as long as it visits and revisits every part of the body. How does this treasure work? In what currency are its transactions conducted? What are its denominations? Such questions would have made no sense to Harvey, for the value of blood was self-evident, and since he regarded it as indivisible the very suggestion that it might have denominations would have seemed absurd. But if you take a spot of normal blood, spread it in a thin film and examine the slide under a microscope, you can immediately see that the fluid has a texture. It is not just a uniform pink substance: there are millions of tiny pink particles. And if you flood the slide with a special stain, the picture immediately springs into sharp relief. The field is crowded with flat pink discs, all of which are the same size, shape and colour. Blood, it seems, is a population, and the redness is confined to the millions of cells that crowd the field from side to side — the featureless, yellow fluid in between is quite empty.

If this film is compared with one taken from a pale, anaemic

Blood as it would have appeared down early microscopes. Without foreknowledge of the existence of blood corpuscles, scientists would have had no good reason for thinking that these were blurred images of a clear-cut reality.
(The microscopes used:
above left   *1700 Tripod,*
above right   *1710 Marshall,*
below left   *1745 Cuff,*
below right   *1760 Martin)*

On the preceding pages
An enlarged smear of a normal blood cell.

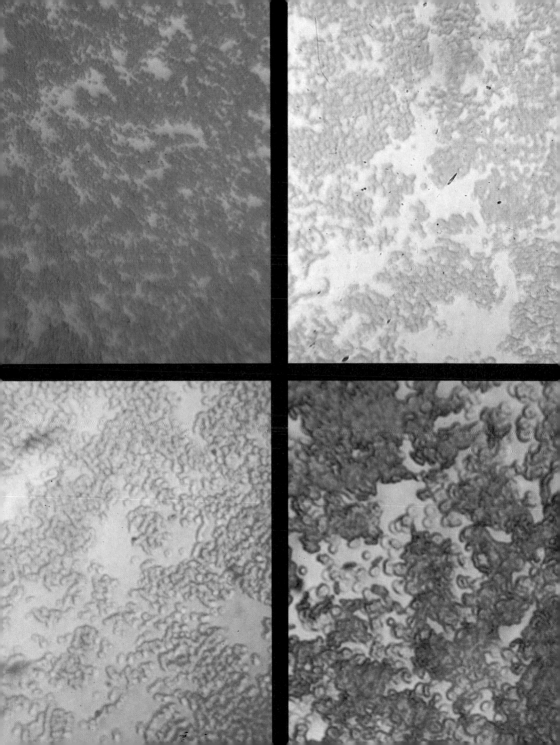

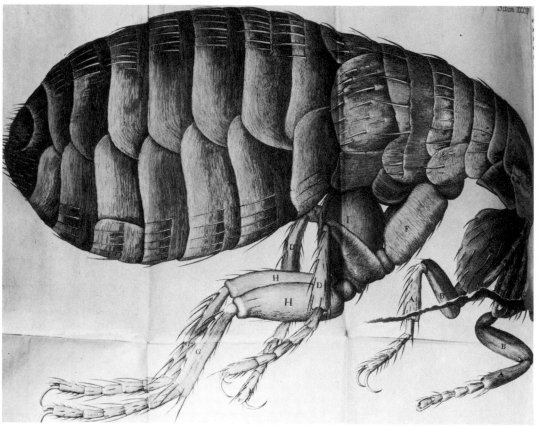

Illustrations from Hooke's 'Micrographia'. The objects represented here (ant and flea) were already recognised and known to have a hard edge. So that even if they had *looked blurred under the microscope there would have been a good reason for regarding the blur as an optical artefact.*

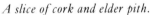

A slice of cork and elder pith.

patient — say, from a pregnant woman who has not been eating the right diet to satisfy the iron requirements of her growing child — certain differences are immediately apparent. All the cells are different sizes, and the colour of each seems to be much less intense than it is in the normal film. Is the paleness of the cells enough to explain the pallor of the patient, or are there also fewer cells than there should be? Is this under-populated blood, or blood with an under-coloured population?

Modern haematologists can distinguish many different types of anaemia: ones where there are fewer red cells than usual, although each cell has a normal size and colour, and ones where each cell is the wrong size and the wrong colour. There are many combinations of error, each one of which has a different cause, and each of which produces a distinct clinical picture. For the haematologist, the anaemic patient is sometimes pale, but invariably interesting.

Such an interest would have mystified the scientists of the seventeenth century. As far as they could see with the naked eye, blood was a simple, homogeneous red fluid — a ruby flood. And even when the newly invented microscope showed that there were millions of particles floating in it, it was more than 200 years before anyone recognised the need to count, measure, classify and name these. Such indifference was partly the result of the poverty of the primitive microscope. Until the middle of the nineteenth century, lenses were so crude that they were incapable of resolving a clear image. The images seen down such early microscopes were so blurred that it would have been impossible to distinguish misleading artefacts from interesting structures.

But it is not simply a question of bad optics. It would be wrong to assume that if a seventeenth-century scientist had been given the opportunity of looking down a twentieth-century microscope he would automatically have discovered the existence of red blood cells — even if he had happened to use that word to mention what he had seen. As it happens, the English scientist Robert Hooke did apply the word 'cell' to the little compartments which he saw when he turned his lenses on to thin slices of cork and elder pith, and, since the appearances which he mentioned do in fact coincide with the entities we now call cells, it could be said that in a sense Hooke

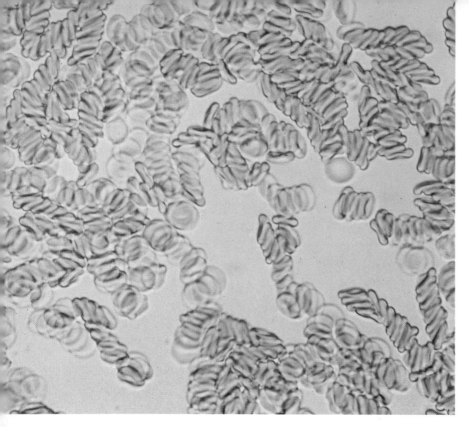

'discovered' them. But there is an equally important sense in which it could be said that he did nothing of the sort.

This would become apparent if we could ask Hooke what description or descriptions he would be willing to substitute for the word 'cell', and then compare his answer with the one that would be given by a twentieth-century scientist in reply to the same question. The descriptions would not coincide, although they would of course overlap, in that both men would probably mention the rectangularity of the little compartments. But for the twentieth-century scientist, the geometrical appearance would be a comparatively unimportant part of the description. For him, the word 'cell' would imply all sorts of notions about the origin and behaviour of the thing which would be conspicuously absent from the description Robert Hooke would be willing to give.

Words or names are not simply straightforward labels stuck on to items of luggage in the natural world, and the mere fact that two people apply the same word to the same visual experience does not mean that they are referring to the same thing. Names are alternatives to an elaborate and often long-winded description of the thing they mention or refer to, and

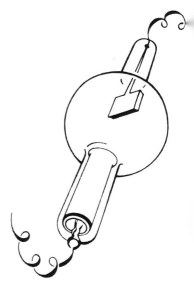

the description which one man has in mind when he uses such a word may be altogether different from the one someone else is thinking of. For a seventeenth-century microscopist, the word he used to describe what he saw down his microscope would refer to its appearance and little more, whereas when a twentieth-century scientist identifies something as a cell, he is recognising what he sees as an instance of something that has properties over and above the ones which can be detected at that particular moment. In fact, his belief in the existence of these properties may be so strong that it actually shapes and alters the appearance of what he sees.

Good visibility, then, may be a helpful condition for making discoveries, but it is neither necessary nor sufficient. If someone already knows what he is looking for, he may recognise it even when the visibility is appalling: but if he has no preconceptions of what to look for, he may misconstrue or completely ignore what he is looking at. An example of the first possibility can be seen on page 221. Although the picture has been deliberately blurred by a computer, one is almost bound to see a familiar face, and although the blur makes the expression unrecognisable, it is almost impossible not to detect the Gioconda smile. Yet to someone who had never seen Leonardo's famous painting neither the face nor the expression would be recognisable; and to someone who had no knowledge of human faces at all, the figure might not even be distinguishable from its background.

The reverse situation is illustrated by the Figure on page 220. No problem of visibility here: the outlines have been deliberately sharpened. And yet to someone who was unfamiliar with the apparatus used in physical laboratories, the clarity of the image would be altogether unhelpful. It might even be hard to tell which way the object was orientated. To see how something is arranged, it is necessary to know what sort of thing it is. And that means knowing what it would look like from all possible angles.

These are some of the difficulties associated with looking at a blood film down a microscope. Unless you were already familiar with the modern notion of cells and all that the word implies, you might not attach any importance to these pink blobs, and unless you had good reason for knowing that they ought to look sharp you might accept their fuzziness as part of their natural appearance. When it came to naming or labelling what you had seen, the word you chose would almost invariably convey theoretical notions about what sort of object it was. For instance, the word 'globule' might be used. But although such a word is much more noncommittal than 'cell', it still has overtones of identification. Globules of what? Strictly speaking, the word means nothing more than 'spherical', but it is usually applied to particular sorts of substances and is not quite as noncommittal as its etymology suggests — by long association the word 'globule' implies fat or oil, so that when it is applied to the pink patches that appear in a blood film, it raises all sorts of associations which may or may not be justified. On the other hand, if you lived in a world which was

Faces and vases compete for recognition in the picture, but only for a spectator who accepts the possibility of either.

entirely made up of perforated sheets or membranes, you would probably see this film as more of the same — as a brilliant white surface punctured with holes through which you could just see vistas of a pink beyond.

Everyone approaches the world with ontological assumptions, confident presuppositions about what there is. According to one philosophical view, the universe is a vast, plural array of nameable things — rocks, lakes, mountains, clouds and people; teacups, clock faces and soup spoons — although other philosophers insist that the universe is really a single, indivisible whole. There is the ancient idea that the universe has nothing in it but basic elements — substances that shape themselves into an infinite variety of transient forms. There are those who insist that even substances are an illusion and that the only knowable things are certain palpable qualities — wetness, dryness, coldness and heat. The most abstract idea of all is the one which says that the only real things in the universe are arithmetical numbers: unity and zero, twoness and trinity, duets and quartets.

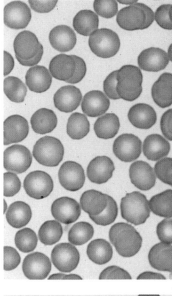

These theories are not necessarily incompatible. The scientists of the early seventeenth century inherited a curious hotchpotch of such ideas from their Greek predecessors, and when they looked at blood they interpreted what they saw and saw what they looked at in the light of these assumptions. Blood to them was a simple, irreducible substance — one of four primary fluids which duplicated in the human body the natural elements of the physical world. For more than 2,000 years, European scientists had been enchanted by the Ancient Greek theory of the four elements which, mixed in various proportions, could produce any of the known forms of physical matter.

This theory combines three of the ontological notions I have just mentioned. It suggests that the world is composed of substances rather than things. It implies that these substances — earth, air, fire and water — express the qualities of heat, cold, wetness and dryness. Fire, for instance, was the elemental expression of hotness and dryness, whereas earth was the result of mixing coldness and dryness. Finally, it celebrates the arithmetical number 4, a digit which has always held a peculiar fascination, recurring in all sorts of familiar, symbolic quartets — seasons, winds, times of the day, points of the compass and ages of man.

Opposite *Normal blood smear. Good lenses and subtle stains produce an intelligible 'blood picture'. Nevertheless, like any other picture, it has to be 'read' as well as viewed. Before he can 'see' what it 'shows' the haematologist has to apply certain conventions of visual interpretation. For example, although the film is thin enough to give the impression of zero depth, the skilled eye 'sees' clues of 3-dimensional structure. The haematologist knows that the red blood cell is a bi-concave disc and that this is why each cell looks paler at the centre. He also knows, or supposes, that his stains will pick out the white cells in purple and that the cells with lobed nuclei are phagocytes, while those with large, round nuclei are lymphocytes. It took time before scientists could reliably assume that the 'appearance' of such artificial preparations corresponded to a natural reality. For all their clarity they could have been seductive artefacts.*

Since the Greeks regarded man as a small replica of the natural world, they assumed that each of the four elements would be represented in the living body. The Hippocratic authorities claimed that there were four physiological fluids or humours: blood, phlegm, and two forms of bile — one yellow and the other black. Blood corresponded to air and expressed the combined qualities of heat and wetness; phlegm duplicated the element of water, since it was cold and wet: yellow bile or choler corresponded to the dry heat of fire, while the

Emblematic representation of the psychological *temperaments* associated with each of the physiological humours. *In the ideal state — in Adam before the Fall — the four humours exactly balanced one another. In ordinary mortals, however, there was always a recognisable surplus of one of them. A modest surplus produced one or other of the four varieties of normal temperament. Great excess led to disease. (Warburg Institute)*

cold dryness of earth was represented by black bile or melancholy. Health depended on the equal proportion of these four humours, although an accurate balance was never found in any living individual — Christian writers assumed that they had never recovered their equal representation in the human body after the Fall of Adam. Slight disproportions produced the familiar variations of the so-called temperaments: a modest excess of yellow bile made a man irritable or choleric; a slight imbalance in favour of black bile caused the melancholy temperament; a surplus of the cool wet humour made a man stolid and phlegmatic. A vast excess of any of the four humours, however, led to recognisable disease.

For obvious reasons, blood was regarded as a privileged member of the quartet, and the sanguine temperament was looked on with special favour. When William Harvey reconsidered the matter in the middle of the seventeeth century, he disregarded the other three humours, insisting that blood alone was the sovereign principle of life:

> It is unquestionable and obvious to sense that blood is the first engendered part when the living principle in the first

A bizarre attempt to personify the attributes of two natural elements, fire and water. (Arcimboldi, Kunsthistorisches Museum, Vienna)

Opposite *Although blood was one of four physiological humours it was regarded as the most important one — the source of life itself. Painters such as van der Weyden recognised this when they represented the Godhead in blood red. (Van der Weyden, Johannesalter, Gemäldegalerie, Berlin)*

226

instance gleams forth. We conclude that blood lives of itself and that it depends in no wise upon any parts of the body. Blood is the cause not only of life in general but also of longer or shorter life, of sleep and of watching, of genius, aptitude and strength. It is the first to live and the last to die.

To some extent, Harvey had arrived at this conclusion as a result of his famous investigations on the developing chick, when he mistook the pulsating embryonic heart he saw for a pounding spot of blood and attributed the pulsation to the blood's innate liveliness. He regarded blood as a pure substance, recognising at the same time that it was imbued with a volatile principle which would evaporate as soon as it was shed:

> When this living principle escapes, the primary substance of blood is forthwith corrupted and resolved into parts. First into cruor or gore … afterwards with red or white parts. And of these parts some are fibrous and tough, others ichorous and serous in which the mass of the coagulum is wont to swim. But these parts have no existence severally in living blood, and it is only in that which has been corrupted and resolved by death that they are encountered.

Harvey's view of blood bears a remarkable resemblance to the eucharistic view of communion wine.

The invention of the microscope a few years later showed that the situation was somewhat more complicated than he had supposed. Blood was seen to have a texture: it had parts; it was several; it was plural. But this discovery did not lead to the recognition of cells, and there is no reason why it should have. Scientists still believed that the body was made out of primary substances, and the fact that one of these substances was now shown to have a texture did not significantly undermine their basic assumption. When Hooke saw pictures like those on page 217, they didn't change his opinion about the constitution of living things. Presumably, it was no more remarkable than finding through a telescope that a wall was made of bricks rather than being a single mass of baked clay — or that a tapestry is composed of separate stitches, or that a mosaic

Renaissance magicians and philosophers such as Robert Fludd visualised the human body as a miniature replica of the universe – a microcosm whose ingredients reproduced and sympathised with the elements and principles of the world at large.

The 'cellularity' of the living body was something more than 'weave', 'texture' or 'grain'. (Opposite, above Unicorn Tapestry, Cluny. Below Detail from the mosaic of St Andrew and St James, S. Prassede, Rome)

228

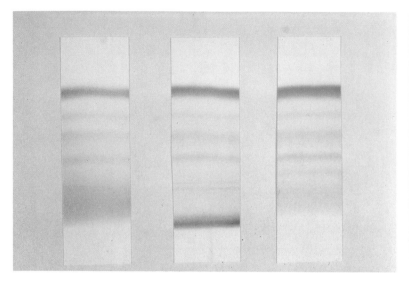

portrait is made of coloured tiles. But when a modern biologist looks at a blood slide, he sees it in an altogether different way. He regards living organisms, man included, as a republic of cells — an orderly multitude of separate living individuals assuming the form of a single person. Oddly enough, less than ten years before Hooke began to use his microscope, the political philosopher Hobbes had represented society as a great Leviathan in which a multitude of individuals acted together as a corporate person. But it was 200 years before this metaphor was usefully applied to the body of the living individual. In 1858 the great German pathologist Virchow declared that 'The structural composition of any living body of considerable size, a so-called individual, always represents an arrangement of a social kind in which a number of individual existences are mutually dependent but in such a way that every element has its special action'.

Such components do not spring into existence spontaneously, nor do they crystallise from a uniform substance. Less than a year after the publication of Darwin's *The Origin of Species*, Virchow announced a fundamental principle which we can usefully regard as the origin of individuals:

> We no longer have any reason to accept the idea that a new cell can arise from a non-cellular substance.
> Whenever a cell arises, there a cell must previously have

Above left   Even the clear plasma has a 'structure'. Not a visible one but one which can be displayed visually, by means of biochemical analysis. These electrophoretic strips show the component plasma proteins, some of which are responsible for carrying immune antibodies.

For Harvey blood disintegrated into 'gore' once it was shed. The picture above would have been evidence of his theory. For a modern haematologist a sample of blood which has settled separates into its natural parts. Red cells sink to the bottom and clear plasma settles on top.

existed. Throughout the living world, plant or animal, whether in part or whole, there is a law of continuous development. All developed tissues can be traced back to a cell.

The objects to be seen on a microscope slide, therefore, are related to one another not like bricks or pots which have been moulded from the same piece of clay and then fired together in the same kiln, but like brothers and sisters who share the same parents and descend from the same ancestors. To all intents and purposes, slides like those on page 224 are family photographs which show in the same frame brothers, sisters and cousins — a generation of living individuals squashed flat into a single commemorative tableau. According to Virchow's principle, each of these cells must have a parent. Where are they to be found?

In order to economise on the use of materials, to keep the mechanics as light as possible, vertebrates have evolved hollow bones with all the structural elements arranged around the edge. The hollow space is occupied by a red marrow, and this is the nursery of the blood. In a spongey maze of blood

Below left   As the red cells descend from the parent stem they slowly mature – the large nucleus directs and dictates the synthesis of haemoglobin, but when the process is finished the nucleus dwindles in order to make room for the oxygen-carrying pigment it has created.

Below right   Frog blood. Amphibian red cells retain their nuclei, whereas the oxygen demands of mammals is so high that in order to maximise the payload of haemoglobin the cells lose their nucleus before they enter the bloodstream.

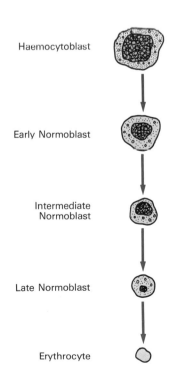

Haemocytoblast

Early Normoblast

Intermediate Normoblast

Late Normoblast

Erythrocyte

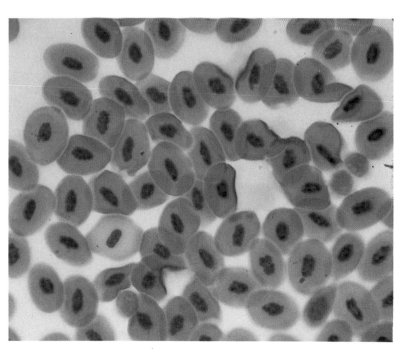

vessels, there are islands and archipelagos of tissue, representing infant blood-cells in various stages of their development. Among these, it is possible to distinguish the so-called 'stem cells', from which the red blood corpuscles take their origin. Each of these stem cells continues to multiply and duplicate offspring, which in turn divide further. But the products are not immediately released into the bloodstream; the offspring have to mature and acquire the correct amount of red pigment, and this process is not complete until several successive divisions have taken place. At each division, the young cells shrink and become progressively redder, and at the moment when they are about to enter the bloodstream they lose their nucleus. They are now ready to be enrolled as fully fledged members of the circulation. Here, they have a short, useful life, which lasts about 120 days. Then, like sputniks in orbit, they run out of fuel, become fragile, drop out of orbit, and are retrieved and scrapped by scavenger cells situated in the liver and the spleen, which strip them there of scarce and valuable components such as iron, which are recycled for the manufacture of fresh cells.

This process may be interrupted or upset at any stage, and each type of interruption produces its own recognisable type of anaemia.

The stem cells may be destroyed by poisons, drugs or radiation, so that the whole process is aborted at the outset. Red cells no longer enter the circulation fast enough to make up for the ceaseless process of exhaustion. As the normal rate of drop-out continues, the blood count falls progressively, and the patient begins to suffer from what is called aplastic anaemia.

Production of red cells may also be held up for want of essential ingredients: iron, protein or vitamins. When iron is in short supply, the cells are characteristically smaller and paler than usual. If the patient is unable to absorb Vitamin B12, he or she develops the characteristic picture of pernicious anaemia, in which the cells are larger than usual and have anomalous shapes.

In the so-called haemolytic anaemias, the cells are destroyed at a much more rapid rate than usual, either because they are congenitally fragile, as in the sickle-cell disease found amongst African Negroes, or because the bloodstream con-

Rudolf Virchow, 1821–1902. Virchow identified the cell as the basic social unit of the living individual, insisting that this was where pathological processes had their seat. Perhaps his political liberalism prompted him to see the body as a biological republic.

Society as a Leviathan of living persons. Hobbes had no reason to suspect that each person was a leviathan of living cells. (Hobbes, 'Leviathan', 1651, frontispiece)

tains destructive substances, the commonest of which are the antibodies found in rhesus incompatability (the RH factor).

Each of these anaemias produces its own characteristic pattern of clinical symptoms. But one duet of symptoms is common to almost all of them, and the consistent recurrence of these two helps to explain one of the most important functions of blood. Seriously anaemic patients are always pale and almost invariably breathless. Why should the colour of blood be associated with the appetite for air?

Towards the end of the nineteenth century — by which time the chemistry of organic substances was undergoing detailed analysis — pathologists successfully identified the pigment contained in the red blood cells as a substance which they called 'haemoglobin'. Within a few years it was demonstrated that this pigment had an unusual capacity for picking up oxygen when exposed to the air and for yielding it up again when exposed to the tissues which required it. The colour of blood, it seemed, was more than a cosmetic feature designed to keep roses in the cheeks: it was a transport mechanism, mediating between the air and the tissues. The red cells were oxygen rafts, and that was the main reason why

Opposite No one will ever have seen an object like this. Atoms have no colour and they are not linked to one another by tangible bonds. Although it may not resemble the object which it represents it immediately conveys all the information the scientists who built it intended.

Right An alternative model of haemoglobin. It accurately reproduces different aspects of the reality it represents.

Spectra produced when solutions of haemoglobin derivatives are viewed with a spectroscope. The right-hand spectrum is of deoxygenated haemoglobin and shows a single dark band in the middle of the spectrum. In the other spectrum of oxygenated haemoglobin, the characteristic change to a two-band spectrum is seen.

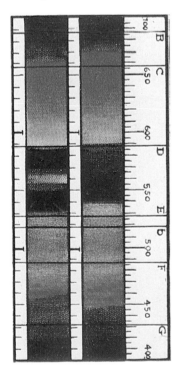

blood had to be kept on the move, had to circulate.

The way in which haemoglobin carried oxygen remained comparatively obscure until physical chemists analysed its structure shortly after the Second World War. The picture on page 234 is a model of a single molecule of haemoglobin. It is not, however, a scale model. No one will ever see anything that remotely resembles this diagram. It is not in any sense a picture; it is not even a sketch. It is a convenient notation which represents and summarises the mathematical results of X-ray analysis.

Although the visible appearance of such a model is an essential part of its function, the significance of its visibility is not quite like that of an ordinary model. In its relation to what it represents, it is like an underground or subway map which does not show the appearance of the tracks or the size of the stations but simply enables one to read off the sequence of stops. Neither the size nor the colour of the little beads convey any information about the character of the atoms they represent. All that matters is that atoms of different elements are unmistakably distinguishable from one another. It is merely a matter of convention that carbon is depicted in black and nitrogen in green — yellow or purple would have done just as well. Similarly, the fact that there are metal rods linking the

235

beads should not be taken to indicate that there are visible lines connecting one atom to the next. The use of such little rods is partly dictated by mechanical convenience — there is no other way of making a cluster of beads stand up — but it is also a conventional way of representing the electro-chemical bonds between adjacent atoms.

Some of the visible features of the model *are* relevant, however, and can be mapped by a knowing eye on to the visual read-out from an X-ray diffraction pattern. From the way in which the plastic 'atoms' are arranged in space, a physical chemist can tell that carbon atom 5 is north-north-east of hydrogen atom 4, and that this in turn is south-south-west of nitrogen atom 96. If the linear dimensions of the rods are taken into consideration, the chemist can also read off the distance between adjacent atoms, converting the centimetres of the model into the Angstrom units of molecular reality.

When a biochemist looks at a model like this, he recognises three distinct chemical components. The main bulk of the model — that tangled blackberry bush — represents a mass of globular protein which serves as a chemical anchorage. In the middle of this thicket is a ring-like arrangement which gives the molecule its characteristic colour. And hanging in the centre of this diaphanous web is an atom of iron, which confers upon the whole arrangement its peculiar affinity for oxygen.

The electro-magnetic properties of this assembly are such that whenever it is saturated with oxygen it appears scarlet. But when it is stripped of its oxygen by the hungry tissues, it absorbs different wavelengths of light and appears purple. This explains Richard Lower's seventeeth-century obser-vations on the colour of asphyxiated blood. It also explains why patients look blue when they are deprived of air for any length of time. When a patient continues to display a bluish tinge even when the atmospheric air is normal, the physician is led to suspect either that his lungs are unable to deliver the amount of oxygen necessary to keep his blood red, or that some structural fault in his heart is allowing venous blood to enter the arteries without first having gone through the north-west passage of the lungs.

When Harvey discovered the circulation of the blood, he regarded the movement as the just and equitable distribution

The extent to which a model resembles what it represents depends on its purpose. These two wooden blocks may not look like the skyscrapers which they represent, and a property dealer would be disappointed if the architect was to present his plans in this form. But the shapes accurately reproduce the aerodynamic characteristics of the projected buildings, and when placed in a wind tunnel they allow the engineer to predict their behaviour and safety.

of a natural treasure. He had no understanding of its underlying value, and, even though his immediate successors showed that the circulation had something to do with the distribution of a substance snatched from the air, it was more than 200 years before anyone realised what this substance was and how it was carried. Now that we know the essential function of oxygen and how it is ferried to and fro by the red corpuscles, it is easy to understand why breathlessness is so frequently associated both with pallor and with blueness. In both cases, the tissues are suffering from lack of oxygen, and in breathing more strenuously the body is automatically trying to compensate for the shortage. If this compensation fails to accomplish its task, the regeneration of ATP — the energy source of the cell — is seriously endangered, and the tissues which

237

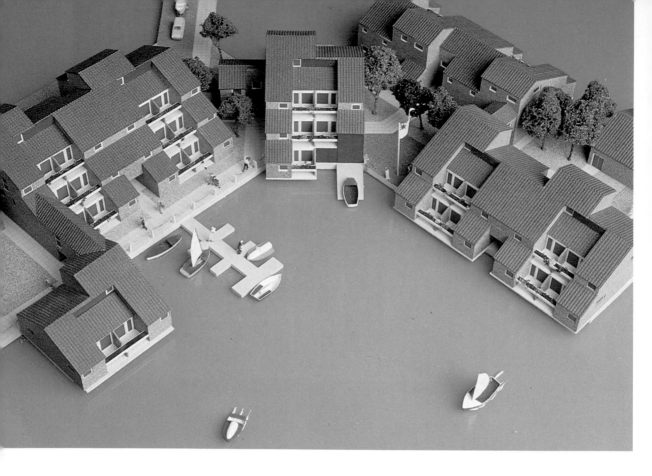

depend upon this energy reserve begin to fail. The patient may start to feel muscular pains or become aware of the exertion necessary to breathe, and in severe cases there may even be angina, as the heart begins to asphyxiate.

At this point a vicious circle is established: as the oxygen shortage becomes more acute, the heart begins to beat faster in the effort to distribute what little oxygen there is. But since these increased efforts also involve the expenditure of energy and, therefore, the demand for oxygen, the whole system starts to pursue a downward spiral. In the face of such deterioration, the compensatory resilience of the body is both defeated and self-defeating, and a fatal outcome can be prevented only by identifying the factor which is responsible. In the case of anaemia, the cause must be identified and corrected — the missing iron or vitamins restored. If the patient is blue, the factors preventing him from being red must be recognised. The management of pathology presupposes the understanding of physiology.

Just as the heart and lungs attempt to compensate for the

In an architect's model the colours are used in a 'digital' fashion to represent the all-or-none differences between different media. Blue represents water, not because the architect wishes to convey the blueness of the lagoon, but because it is a conventional way of distinguishing water from dry land. Primrose yellow would have done just as well.

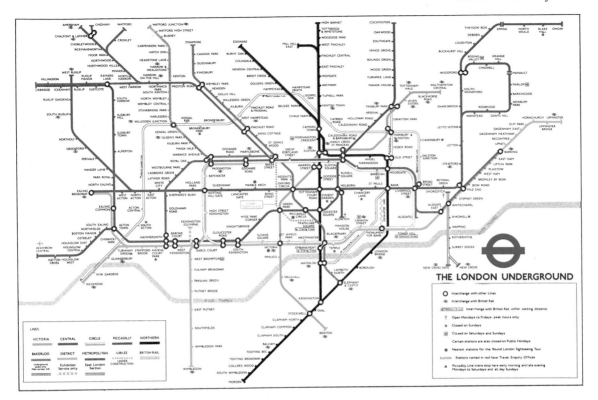

Henry C. Beck's famous map of the London Underground system is not a picture of the railway. It is a colour-coded guide to travelling. It tells the passenger the order in which stations follow one another and where different 'lines' intersect. It conveys no information about the distance between stops, the curve of the lines or the absolute length of the journey.

The colours used in these models are not conventional. They are meant to reproduce and convey the colours of the actual uniforms.

failure of the blood, healthy blood can compensate for failures of the heart and lungs. When lung disease frustrates the intake of oxygen, the bone marrow, operating on the same homeostatic principle as the respiratory centre in the brain, makes up for the shortage by multiplying the birthrate of new red cells and sending them into circulation to snatch whatever oxygen is available. Patients with chronic bronchitis and emphysema, for example, are often found to have a raised red-cell count: there may be as many as 8 million cells per cubic millimetre, as opposed to the 5 million found in normal blood.

A shortage of atmospheric oxygen produces the same effect. At high altitudes, respiration automatically deepens in the attempt to compensate for the smaller amount of oxygen contained in each normal breath. This is, however, a relatively inefficient adaptation: the subject is usually distressed by the strenuousness of his own breathing, and the extra muscular effort consumes so much oxygen that it tends to defeat its own purpose. If the subject stays in such a thin atmosphere for more than a few days, the body must undertake a more rewarding form of acclimatisation. After a few weeks, the blood therefore begins to show a recognisable increase in the number of circulating red cells, and by the end of six months the count may be as high as 10 million cells per cubic millimetre. The extra haemoglobin which is now circulating through the lungs guarantees an effective uptake of what little oxygen there is, which relieves the respiratory centre of the full burden of compensation. This is why a new arrival in the Andes is prostrated by breathlessness while the local inhabitants are able to do heavy work and even play football without any sign of respiratory distress.

Such a high blood count is an expression of furiously energetic activity on the part of the bone marrow. But since the extra output is superimposed on a comparatively high level of normal activity, the achievement is not quite as impressive as it seems. The blood is constantly renewing and turning over the circulating population of red cells. Each cubic millimetre loses more than 20,000 of its 5 million cells every twenty-four hours. The output needed to compensate for such wastage is so enormous that it takes only a comparatively small bulge in the birthrate to raise the count by the amount I have just mentioned. As with so many of the

The tailor's dummy 'models' the client by reproducing her metrical dimensions. It does not reproduce her appearance but it guarantees that her appearance in the new garment will be presentable.

This model reproduces the appearance *of the object it* represents in every important aspect except size. *(Left* Lord Hunt *with a model of Everest)*

Living organisms cheat death by restocking the world with fertile replicas of their parents. The process preserves a nice balance between variation and uniformity. The individuals of given species differ from one another, but not as much as they collectively differ from the individuals of a closely related species.

Opposite   (Rembrandt, details from self-portraits: above (left) Alte Pinakothek, Munich, (centre) Mauritshuis, The Hague, (right) National Gallery, London; centre (left) Mauritshuis, The Hague, (centre) Kunsthistorisches Museum, Vienna, (right) National Gallery of Art, Washington; below (left) Frick Collection, New York, (centre) Iveagh Bequest, Kenwood, (right) 'The Artist's son, Titus', Dulwich College Picture Gallery, London)

THE BODY IN QUESTION

processes by which the body adjusts itself to occasional threats and stresses, the activity is no more than an amplification of the one by which normal function is maintained. Like the skin and the intestine and all the other tissues whose cells retain and exercise the ability to go on dividing, the bone marrow continues to grow throughout adult life. But since the chronic wear and tear keeps pace with the rate of multiplication, the growth does not result in an increase in size. Nevertheless the process is identical with the one by which the creature does enlarge and achieve its mature form. So in a very important sense the maintenance of the structure is simply a continuation of the method by which that structure first came into being. Maturity is the point at which a balance is struck between growth and decay.

This immediately raises the question: why is the creature not immortal? If so many of its tissues are capable of renewing themselves, and if by amplifying these processes they are capable of repairing all but the most overwhelming injuries, why does life not continue indefinitely? There are two possible answers to this question. First, not all tissues do retain the capacity for unlimited growth; the cells of the nervous system, for example, cease to divide once the mature structure has been achieved, and whenever cells reach this stage they begin to accumulate uncorrectable metabolic errors. Although the body can accommodate itself to a certain number of these failures, there comes a point when the system as a whole is so incoherent that the homeostatic capacities begin to falter, and the creature becomes fatally inefficient. Second, the retention of cell division is not quite such an effective guarantee of immortality as it was once thought to be. Biologists such as Alexis Carrel in the earlier part of this century originally claimed that any cells which retained the capacity to go on dividing were potentially immortal and any tissue which was made up of such cells automatically renewed its vitality by multiplying fresh offspring.

This is now known to be untrue. There appears to be an upper limit to the number of times any cell can reproduce itself, and modern biologists are beginning to suspect that senescence is built into the biochemical instructions of the living cell, and that the ability to divide and reproduce is just as susceptible to metabolic error as all the other functions. If this is so, we must regard death as an intrinsic feature of life and not merely an unavoidable interruption of it.

But our reappraisal of the situation may have to be even more radical than this. We tend to think that birth and reproduction are nature's way of overcoming death, whereas it may be that death is nature's way of giving free rein to the creative opportunities offered by birth. If all creatures were immortal and preserved the same functions and abilities throughout their endless lives, they would pre-empt the possibility of improvement or adaptation. Sexual reproduction, however, gives the species the recurring opportunity of bringing together unprecedented combinations of genetic material, and, with the further novelties which are added by mutation, the species can audition untried newcomers. In the absence of senescence and death, these young hopefuls would enter an overcrowded stage. By modest withdrawal from the scene one generation acknowledges and gives way to the unforeseeable talents of the next.

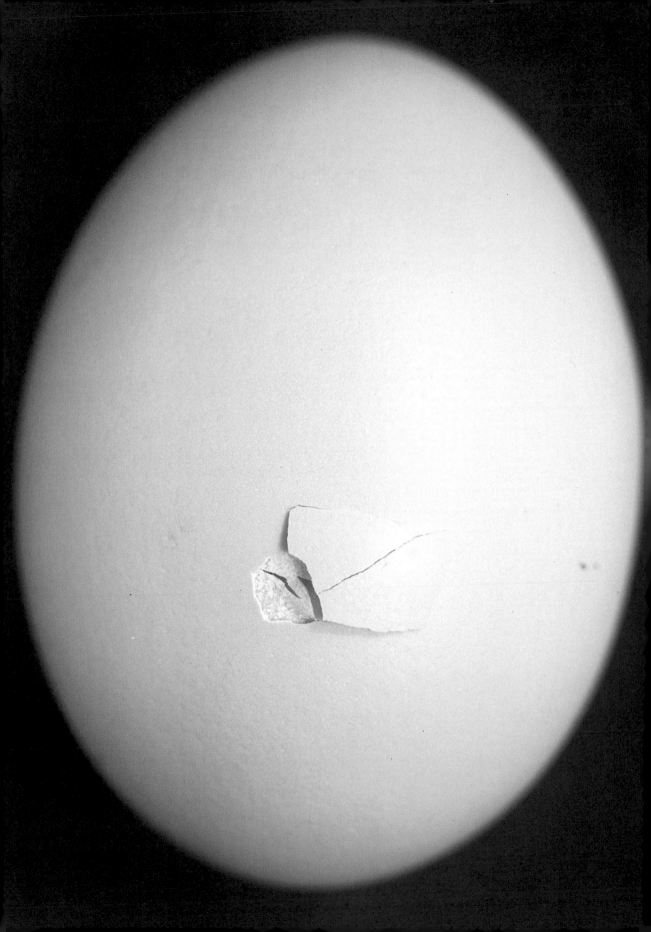

THE IDEA OF CIRCULAR RENEWAL IS ONE OF THE OLDEST themes in European culture. Men had only to open their eyes to see that the sun and moon rose and fell beneath the horizon, appearing, disappearing and reappearing every twenty-four hours. The heavens revolved in their vault, and the seasons repeated themselves; the clouds poured their rain on to the earth, which then gave it back to the heavens in the form of vapour. Nature spent and replenished herself with majestic regularity. These cycles seemed so universal that poets and philosophers came to regard them as outward signs of some deep metaphysical principle of giving, receiving and returning, which applied not only to nature but to human morality as well. The popular image of the Three Graces, for instance, personifies the recurrent phases of generosity: one gives, another receives, while the last returns her gift to the original donor — a circular entanglement of charity.

On the preceding pages *Two of the stages in the hatching of a chick.*

Since the ancients regarded the human body as a miniature replica of the cosmos, it seemed inevitable that they would find counterparts of this rhythmic circularity in all physiological processes. Hippocrates and Plato visualised the movement of the blood as something already prefigured in the drift of stars. And although we now think of William Harvey as the first exponent of scientific physiology, he was also inspired, as Newton was, by the metaphysical doctrines of antiquity. His physiological researches were directed by mechanical considerations, but in demonstrating and proving the existence of a circulation, he at the same time satisfied his almost religious commitment to the occult significance of the circle. It is not surprising, therefore, that in later life he turned to one of the other great rotations of nature: the endless cycle of growth, death and reproduction:

> In this reciprocal interchange of Generation and Corruption consists the Eternity and Duration of mortal creatures. And as the Rising and Setting of the Sun, doth by continued revolutions complete and perfect Time; so doth the alternate vicissitude of Individuums, by a constant repetition of the same species, perpetuate the continuance of fading things.

Like his predecessors, Harvey was struck by the fact that individuals earned their immortality by proxy, perpetuating

248

*The reciprocal circularity of
the elements.*

their otherwise impermanent form by handing it on to the
generations that followed, and that this process repeated itself
with orbital regularity. Unfortunately, the link between successive generations was invisible, and there was no mechanical
metaphor with which to bridge the gap. In order to explain
how the reproductive mystery fulfilled itself, Harvey reverted
to the ancient doctrine of the shaping supremacy of the soul,
an idea which he had inherited from Aristotle, who was
himself expressing one of the most universal intuitions of the
human mind. Living in a body that acts so faithfully in
obedience to his will, man has found it almost impossible to
shake off the conviction that the changes of the physical
universe are the outcome of mental processes like his own and
that any alteration in the state of things is the expression of
agency as opposed to causality; in short, that all events are
actions.

Aristotle saw the universe as a creative partnership between
mind and matter. Matter, an inert compound of four primeval
elements, was shaped and ordered by an insubstantial power
which could imagine and bring into material existence an

249

Left *The Roman Stoic, Seneca, saw the Three Graces as an allegory of liberality in which the linked figures enact the successive stages of giving, receiving and returning. (Botticelli, 'La Primavera', detail, Uffizi, Florence)*

Right *In his enigmatic* Allegory of Prudence *the ageing Titian represented the passage of time and 'the continuance of fading things'. The art historian Panofsky suggests that this picture may have been used to decorate the cupboard in which the artist deposited his will. (National Gallery, London)*

infinite variety of forms or ideas. As in the Old Testament, a spirit brooding on all possible forms moved over the surface of the shapeless matter and drew it out into the familiar beings of nature.

In his great work *On the Generation of the Animals* in the fourth century BC, Aristotle made it clear that this cosmological process was repeated on a small scale in the conception and development of living organisms: the male sperm was the spiritual agency which conjured limbs and organs from the menstrual blood provided by the female — just as mountains, rivers and continents had been moulded from formless matter by the soul of the universe. Embryology and cosmology mirrored each other: both were examples of inventive handicraft. At this stage in human thought, art and craft were the only available metaphors for making the experience of transformation understandable. The most intelligible instances of change were those wrought by man himself, who cooked, brewed, baked, wove, hammered, sawed and smelted his ideas

Above   *Illustrations from Rueff's 'De Conceptu et Generatione Hominis', showing the principle of epigenetic development. The foetal image gradually emerges from a formless mass.*

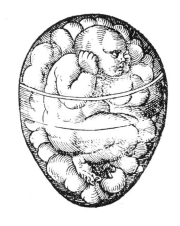

into physical existence, quarrying the inert materials of the physical universe and giving body to his concepts.

Aristotle was so strongly committed to the idea of the sperm's creative abilities that he regarded the female contribution as being altogether amorphous: the male principle was the shaping influence under which the foetus gradually assumed its form. Some of his predecessors had claimed that certain foetal parts were already in existence at the time of conception, but Aristotle refused to accept the idea that development was achieved by assembling prefabricated elements. For him, growth and development were 'epigenetic' processes, which meant that all parts of the foetus emerged and became recognisable at the same rate. There was no pre-existent structure. For Aristotle, physical existence had two complementary aspects: structureless substance and ideal form. By lending its plastic possibilities to the shaping power of Ideas, the universe built up its inventory of recognisable things.

The object made by a craftsman realises an idea which must be held before the mind's eye throughout the task. The shaping hand is continuously guided by a visual image.

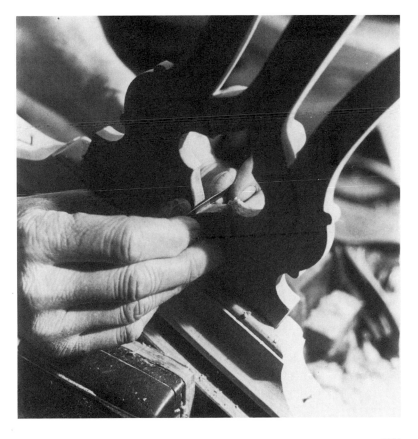

The fertilising power of 'soul' or 'spirit' remained an enduring theme in European thought and the Christian notion of immaculate conception gave it peculiar emphasis. It is not altogether surprising, therefore, that when William Harvey applied himself to the problem of generation, he adopted Aristotle's belief that the active principle in development was essentially psychic and even drew a parallel between the womb and the brain, pointing out that both were organs of conception. However, whereas Aristotle regarded the female contribution as totally inert and thought that the foetal shape was conferred upon it by the semen, Harvey considered the ovum the source of potency and the sperm simply the means of provoking it into action, thereby releasing its inherent potential for growth and development. Both Aristotle and Harvey denied the existence of preformed parts and insisted that differentiation took place under the influence of a spiritual agency. Like so many theories which rely on mystical principles, these proposals are unacceptable not because they are *wrong*, but because there is no intelligible way in which they can be tested, which means that there is no way in which they can be *shown* to be wrong.

Since the theory fails to account for the normal process of development it doesn't even begin to explain the occurrence of congenital abnormality. Or rather, it explains it too easily, for once development is attributed to an imaginative process anything which is mentally conceivable is biologically possible.

Perhaps this is the reason why bizarre monsters and imaginary beings figure so prominently in early treatises on natural history, and why at the same time there is such an inadequate report of genuine congenital abnormalities. It is only too easy to compose a fabulous bestiary by shuffling and transposing the various parts of familiar animals, thus creating a colourful card pack featuring griffins, hippogryphs and cameleopards. Such creatures are listed without critical comment alongside real animals, as if the mere fact that they could be imagined were a guarantee of their probable existence. Sometimes the monsters were created by reshaping the proportions of the human frame. Travellers' tales include reports of men with ears as large as palm fronds, sciapods who shaded themselves from the tropical sun with their one enormous

The sleep of reason produces monsters. But the prodigies created by the human imagination have a different type of appearance from the abnormalities which result from faulty embryological development.

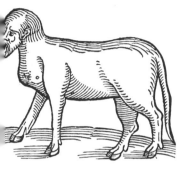

foot, antipodes whose legs faced backwards and men whose heads grew beneath their shoulders. The problem is that the imagination can play only with the repertoire of normal forms, whereas the way in which congenital abnormality departs from the ordinary could never be anticipated by the untutored imagination. So although there are occasional reports of babies born with six fingers or two heads — aberrations which might be imagined by someone who had never seen them — the monotonous regularity of examples such as harelip and spina bifida could easily escape notice for the simple reason that they are not the sort of deformities which would be spontaneously imagined.

As a general rule, human beings tend to overlook or at least misrepresent the appearance of anything which doesn't already figure in a familiar and well-established classification. And since there was no conceptual interest in the various fissures and clefts which occasionally marred the newborn

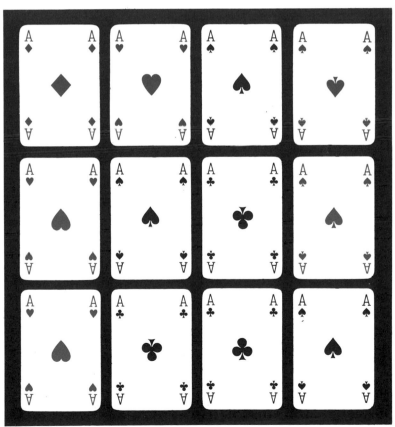

It is easy to overlook anomalies if they don't conform to a conventional stereotype.

child, such examples are understandably conspicuous by their absence in the early treatises on human natural history. Only when interest in the exorbitant gave way to a concern for the regular and the ordinary did a genuine interest in congenital abnormality begin.

Regrettably, interest in the bizarre and the improbable has resurfaced in our own time — less excusably now, since it coincides with an unprecedented growth in scientific understanding. For some reason, many people turn their backs on the majestic regularity of what has been discovered and lend their belief to the limitless possibilities of what can be imagined. The spoon-bending forces of the mind take precedence over the physical forces which straighten the conduct of the atom. Close encounters of the third kind preclude the more modest transactions of the first and second.

By the end of the seventeeth century, the rising prestige of so-called mechanical philosophy made scientists reluctant to

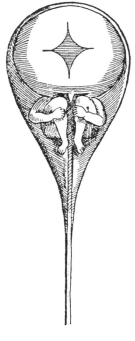

Hartsoeker's drawing of a human spermatozoon.

Opposite   *The preformationist theory of development.*

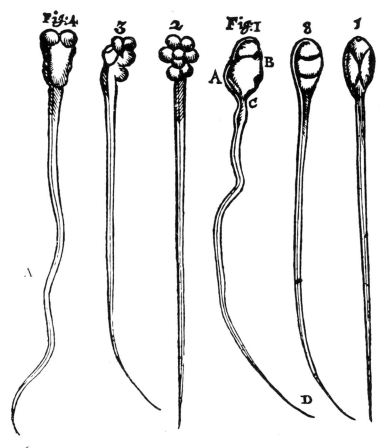

Left   *Spermatozoa were seen down the microscope in the late seventeenth century but the full significance of fertilisation was not recognised until two hundred years later.*

256

accept explanations based on the continuous action of psychic powers. Everything was done to exclude mental urges, and before long this attitude began to influence theories of generation and development. Many biologists reverted to the ancient doctrine of preformation, insisting that the primordial germ already contained a miniature replica of the adult organism, and since this had only to enlarge and unfold itself there was no need to invoke a supervising craftsmanship. Enlargement and unfurling could easily be interpreted in mechanical terms. All that the foetus had to do was to absorb and appropriate the physical nourishment which surrounded it in the womb or the egg.

This still left the problem of where the original effigy arose, and opinion divided itself just as it did in the case of the epigenesists. Some thought that the *petite icône* was inherited from the father, others that it was a maternal bequest. With Leeuenhoeks's discovery of microscopic spermatozoa in 1677, the controversy between the ovists and animalculists became more intense than ever. Some biologists even managed to persuade themselves that they could see the preformed miniature huddled like an astronaut in the forward hatch of the sperm. And when their opponents pointed out that there was no such creature to be seen, the preformationists declared that it was transparent in the early stages of development and was therefore bound to escape detection.

The scientists who took preformation to its logical conclusion committed themselves to an infinite regress of miniature forms, packed inside one another like a nest of Russian dolls. If all the details of the adult form were represented in the miniature effigy contained either in the egg or in the sperm, it was necessary to assume that the germ cells of the adult were already included in the foetus. These in turn presumably prefigured the unborn adults of the generation to follow, and so on, back to the first moment of recorded time.

This arrangement made continuous creation unnecessary, but it did so only by attributing the whole concentric design to a single burst of creative activity on the part of God. The theory of *emboîtement*, as it was called, implied that the details of posterity were already prefigured in the first occupants of Eden, and some scientists even went to the lengths of calculating the number of individuals represented in miniature in the womb of Eve. Scientists who were unable to tolerate the continuous creativity of an anonymous craftsman found it easy to accept an abrupt and self-limiting outburst of craftsmanship on the part of Providence.

What made this theory attractive was the fact that it reflected the mechanistic Deism of the early eighteenth century. In physics and astronomy scientists were trying to avoid explanations which invoked the action of spirits or souls, but the only way they could reconcile this with their Christianity was to suppose that God's creative influence was limited to the dawn of time and that, having manufactured and wound up his cosmic clockwork, he abstained from all further action and amused himself by listening to its regular chimes. Preformationism was simply an extension of the same idea.

There were, however, some awkward drawbacks to this theory. It made no allowance for the unexpected occurrence of congenital abnormalities — unless one were to assume that God had mischievously included an occasional grotesque. Detailed preformation also made it hard to explain the well-recognised phenomenon of healing and regeneration. If everything were preformed, it was impossible to account for the growth and replacement of lost parts unless one made the baroque assumption that all future accidents were anticipated in the design and that each potentially losable part was backed up by an infinite regression of replacements. The most serious objection, however, was that the theory made no intelligible allowance for the way in which offspring often bore such striking resemblances to both parents and that family features were inherited and distributed in a piecemeal fashion.

The preformationists could not provide satisfactory answers to these objections as long as they remained committed to a pictorial or iconic view of the developmental process. Admittedly, the only way of avoiding a psychic agency was to postulate a material predeterminant — some structure whose previous existence in the fertilised egg controlled and directed the developing outcome. What they failed to see was that the outcome of a process can be physically predetermined without having to be visually prefigured.

We now know that the specifications which control the manufacture of an object need not resemble the end-product. When a modern engineer wants to reproduce an elaborate

One way of reproducing an image is to 'pull' a print of it from an engraved plate. The print is exactly the same size as the plate...

... Nature obviously employs a different technique.

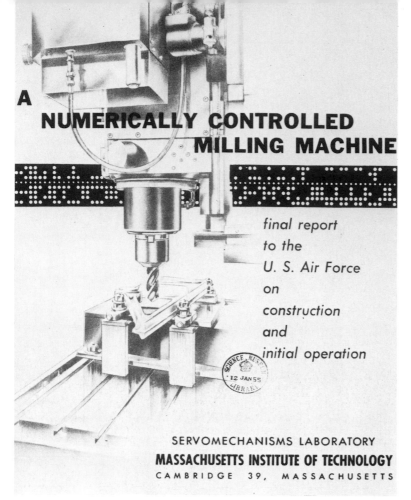

A
**NUMERICALLY CONTROLLED
MILLING MACHINE**

final report
to the
U. S. Air Force
on
construction
and
initial operation

SERVOMECHANISMS LABORATORY
MASSACHUSETTS INSTITUTE OF TECHNOLOGY
CAMBRIDGE 39, MASSACHUSETTS

*The pattern of perforations
represents a sequence of
instructions which the machine
'reads' and 'obeys' without
having to conjure up a visual
image of what it is making.*

component, he does not have to draw a diagram of it: he has
only to write a set of peremptory commands, and as long as
these are obeyed in the order in which they are printed, the
object will automatically take the required shape. The instructions can be fed directly into a machine in the form of a
punch-tape, where the pattern of perforations dictates the
movements of the cutting edge. The machine effectively
translates the instructions into the movements needed to
bring about the desired shape. When a television picture is
transmitted, there are no ghostly images flying through the
air: the picture is converted into a series of visually unrecognisable radio pulses, and when these are picked up by the
domestic aerial, they cause variable deflections of a flying spot
on the television screen and thus reconstitute the original
picture. In both these cases — and there are many other
examples — a linear code dictates the construction of a visible
object.

A linear programme which dictates a continuous line of varying brightness can yield a spatial picture.
Above   *The spiral scan of an engraving. (Detail from 'The Napkin of St Veronica', engraved by Claude Mellan, 1649)*

Below   *The zig-zag scan of a television screen.*

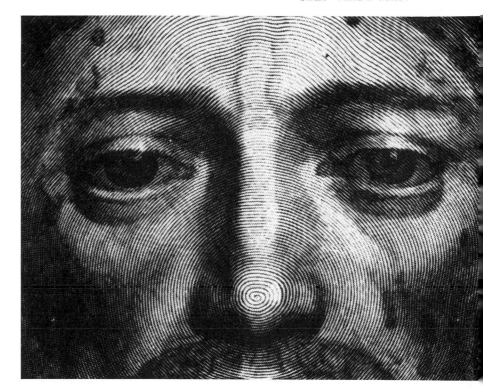

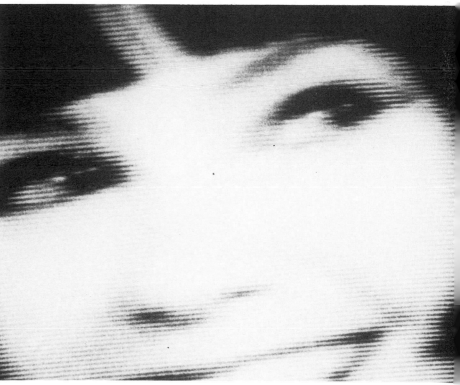

The twentieth century has exploited this principle in almost every field of technology, but the notion had been intuitively recognised and in some cases explicitly understood for many years. Descartes appreciated that the written signs which conjure up pictures in the mind's eye do not resemble the images which they provoke. Musical notation bears witness to the same principle. By decoding and obeying a line of conventional signs, a competent musician can recreate a melody without ever having heard it. Musical boxes simply automate this process. When Mozart wrote rondos for jewelled automata, he did not have to hum the tunes to the mechanical pianist: the composition was translated into a pattern of prickles mounted on a revolving cylinder. As this turned, a comb made of metal teeth was plucked in the appropriate order, and the melody was reconstituted. In the early nineteenth century, the same idea was applied for the first time to an industrial process. The French engineer Jacquard invented a loom which wove elaborate patterns by automatically following a linear programme of instructions stamped on to hinged cards. Although the holes are arranged in a pattern that bears no visual resemblance to the pictures on

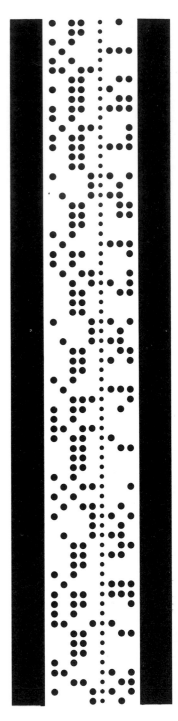

Above *Instruction versus illustration. A sequence of coded factors can dictate a visual outcome.*

Opposite left   When a skilled musician reads the score he may 'hear' the melody without having to perform it. But he can also perform it without 'hearing' the tune, by treating the dots as orders issued to his fingers. (J. S. Bach manuscript)

Right   Instructions like these are often called 'patterns' – but a knitting pattern does not have an 'appearance' – or at least its appearance is not an essential part of its function. It would work just as well if it was read out loud.
Below   A pianola roll and a musical box.

EXPLANATION OF TERMS:—P. Plain knitting.—B. Purl.—T. Knit 2 together.—O. Make a stitch.—A. Knit 3 together.

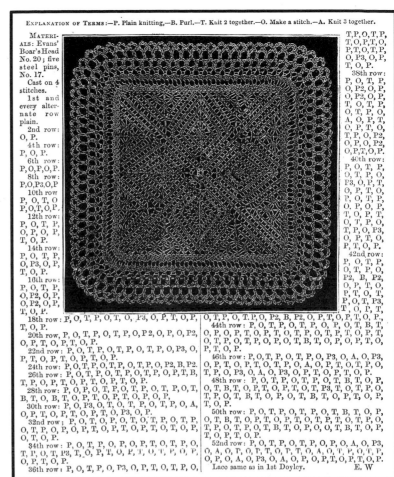

MATERIALS: Evans' Boar's Head No. 20; five steel pins, No. 17.
Cast on 4 stitches.
1st and every alternate row plain.
2nd row: O, P.
4th row: P, O, P.
6th row: P, O, P, O, P.
8th row: P, O, P3, O, P
10th row P, O, T, O P, O, T, O, P.
12th row: P, O, T, P, O, P, O, P, T, O, P.
14th row: P, O, T, P, O, P3, O, P, T, O, P.
16th row: P, O, T, P, O, P2, O, P, O, P2, O, P, T, O, P.
18th row: P, O, T, P, O, T, O, P3, O, P, T, O, P, T, O, P.
20th row, P, O, T, P, O, T, P, O, P2, O, P, O, P2, O, P, T, O, P, T, O, P.
22nd row: P, O, T, P, O, T, P, O, T, P, O, P3, O, P, T, O, P, T, O, P, T, O, P.
24th row: P, O, T, P, O, T, P, O, T, P, O, P2, B, P2, T, P, O, T, P, O, T, P, O, T, P, O, P.
26th row: P, O, T, P, O, T, P, O, T, P, O, P, T, B, T, P, O, T, P, O, T, P, O, T, P, O, P.
28th row: P, O, P, O, T, P, O, T, P, O, T, P, O, T, B, T, O, B, T, O, P, T, O, P, T, O, P, O, P, O, P.
30th row: P, O, P3, O, T, O, T, P, O, T, P, O, A, O, P, T, O, P, T, O, P, O, P3, O, P.
32nd row; P, O, T, O, P, O, T, O, T, P, O, T, P, O, T, P, O, P, O, P, T, O, P, T, O, P, T, O, T, O, P, O, T, O, P.
34th row: P, O, T, P, O, P, O, P, T, O, T, P, O, T, P, O, T, P3, T, O, P, T, O, P, T, O, T, P, O, P, P, O, P, T, O, P.
36th row: P, O, T, P, O, P3, O, P, T, O, T, P, O, ...

O, T, P, O, T, P, O, P2, B, P2, O, P, T, O, P, T, O, P,
44th row: P, O, T, P, O, T, P, O, T, P, O, T, B, T, O, P, O, P, O, T, P, O, T, P, O, T, P, O, T, B, T, O, P, T, P, O, T, P, O, T, B, T, O, P, O, P, T, O, P, T, O, P.
P, T, O, P.
46th row: P, O, T, P, O, T, P, O, P3, O, A, O, P3, O, P, T, O, P, T, O, P, O, A, O, P3, O, P, T, O, P, T, O, P, T, P, O, P3, O, A, O, P3, O, P, T, O, P, T, O, P.
48th row: P, O, T, P, O, T, P, O, T, P3, T, O, T, P, O, T, B, T, O, P, T, O, P, O, T, P3, T, O, T, P, O, T, P, O, T, B, T, O, P, O, T, B, T, O, P, T, O, P.
T, O, P.
50th row: P, O, T, P, O, T, P, O, T, B, T, O, P, O, T, B, T, O, P, T, O, P, O, T, P, O, T, B, T, O, P, O, O, T, B, T, O, P, T, P, O, T, B, T, O, P, O, O, T, B, T, O, P, T, O, P, T, O, P.
52nd row: P, O, T, P, O, T, P, O, P, O, A, O, P3, O, A, O, T, O, T, P, O, T, P, O, T, P, O, T, P, O, T, P, O, T, P, O, T, P, O, A, O, P3, O, A, O, P, O, P, T, O, P, T, O, P.
Lace same as in 1st Doyley.   E. W

T, P, O, T, P, T, O, P, T, O, P, T, O, T, P, T, O, T, O, P3, O, P, T, O, P.
38th row: P, O, T, P, O, P2, O, P, O, P2, O, P, T, O, P, T, O, P, T, O, T, P, O, T, P, O, T, P, O, P2, T, P, O, P2, T, P, O, P2, O, P, O, P, P2, O, P, T, O, P.
40th row: P, O, T, P, O, T, P, O, P3, O, P, T, O, P, T, O, P, O, P, O, P, O, T, P, O, T, P, O, P3, O, P, T, O, P.
42nd row: P, O, T, P, O, P2, B, P2, O, P, T, O, P, T, O, P, T, P3, T, O, P,

the carpet, their linear sequence determines the appropriate movement of the shuttles.

In all these processes, the notion of information takes the place of illustration. The performance is guided by instruction rather than by example. This principle presupposes the existence of a formal or conventional notation whose characters can be combined in unlimited variety. It is not even necessary to have a large number of characters. Racine and Shakespeare were able to distinguish themselves from each other using a repertory of twenty-six alphabetical symbols. Schönberg was able to make an unprecedented contribution to musical form without having to break out of the musical notation which was available to Bach. The advantage of such an arrangement is that it allows one to vary the performance character by character, whereas a system which depends on pictorial example is committed to the indivisible entirety of the pre-existing model.

The biological relevance of this principle did not become apparent until the end of the nineteenth century, partly because it was not widely recognised as a principle, for technologists often succeed in creating profitable mechanisms without appreciating the abstract idea which they exemplify. But even if eighteenth-century biologists had identified the principle of linear programming, they could not have applied it usefully to the problem of heredity. What they lacked was a knowledge of, first, the notation in which such instructions might be written, second, a site which would bear the inscription, and, third, the intervening processes which would translate and execute the programme.

When the word tissue was first applied to living matter scientists were drawing an analogy with textiles. *Since Virchow it was recognised that tissues were* communities *and that they were convened rather than woven.*
Opposite *Tissue sections from a healthy body.*

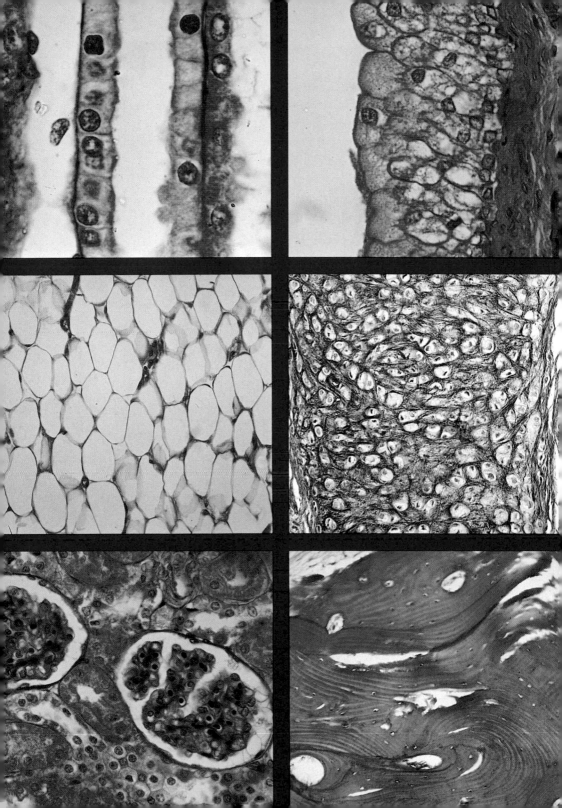

From what was then known about the nature of living matter, creative craftsmanship and preformation were the only conceivable ways in which form could be imposed upon substance. Pure matter was structureless; whereas form was insubstantial. Until some intervening level of organisation could be visualised, the controversy between preformation and epigenesis was bound to remain unresolved.

With the discovery of the cell, biologists identified in one and the same entity the site of the instructions and the agent which executed them. Until the end of the seventeenth century, the various parts of the body were regarded as gross compositions, varying only in consistency and texture. Anatomists recognised that certain organs were spongey and others firm and gelatinous; some were fibrous and closely woven, and others dense and rigid. There were elastic strings and glistening membranes; there were bags, balls and tubes. But, as the microscope became more widely used, it was apparent that these simple distinctions could be broken down still further, and by the end of the eighteenth century biologists were confident that they had found a fundamental unit common to all the tissues of the body — a simple, irreducible element which could be put together in various ways to produce all the known textures. This unit was the fibre, a microscopic thread which could be woven into loose meshes or dense, impermeable sheets, tightly bound into tendons or loosely bundled to form muscles. By 1800 the body was seen as an elaborate textile, a garment of hemps, worsted and linens — the fact that the term 'tissue' was introduced at that time indicates the persuasiveness of the metaphor.

The resolving power of the microscope continued to improve and biologists discovered that there was yet another level of organisation, a hidden arrangement of living matter which turned out to have a different biological status from that of the fibre; something more important than mere grain or texture. They became aware that all tissues were made up of assemblies of small, globular masses: each one anatomically distinct, each one housing a darker centre or nucleus. It soon became apparent that the globules or cells were not decoratively embroidered on to some underlying fabric like buttons or sequins, but that they actually *were* the fabric and that it was the shape, character and arrangement of these

266

entities which gave each organ and tissue its characteristic consistency and function.

During the 1830s there was intense controversy about the origin and character of these structures. Some biologists insisted that they were deposited or crystallised *in situ* and that they were preceded by a primal slime or blastema. Further research showed that the cells which constituted an individual were biological beings in their own right and that no cell could exist unless it had divided or split off from one like it. Taken at face value, a cell could have been visualised as a geometrical alternative to the fibre — a structural component which happened to assume a globular rather than a linear form. For scientists such as Virchow, however, the way in which cells arose and composed themselves indicated that the living organism was a republic of biological persons, dividing their labours and differentiating their functions in order to serve the living commonwealth which they constituted.

It is often pointed out that Virchow arrived at this conclusion under the influence of his political beliefs, and that as a passionate liberal of the 1848 generation he was predisposed to see the organism in republican terms. Social metaphors often recur in biological theory, but, unlike analogies drawn from technology, they do not have immediately testable implications. Harvey may have been stirred but was unable directly to profit from his view of the heart as a monarch. Nevertheless such metaphors occasionally prove helpful, if only for the way in which they redirect scientific attention and free the imagination from even less helpful images. Virchow's republican metaphor recast the problem of reproduction and growth. What was handed on from one generation to the next was neither a picture nor a shaping idea, but a fugitive delegate from the parent assembly bearing all the instructions necessary to convene a new assembly in the constitutional image of its predecessor.

With the direct microscopic observation of fertilisation at the end of the nineteenth century, it became apparent that there were two parental delegates, that whatever instructions were needed to reconstitute the next assembly were contributed by both male and female and without such collusion the biological charter would be incomplete. These descriptions went a long way towards satisfying the three conditions

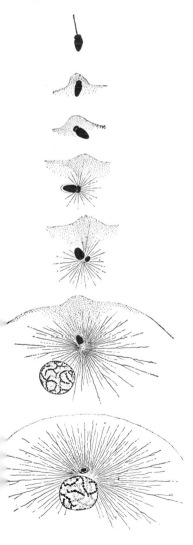

The sperm head contains the paternal instructions. One of many competitors succeeds in penetrating the membrane of the ovum. Two sets of instructions are now in place and the first cell division can begin.

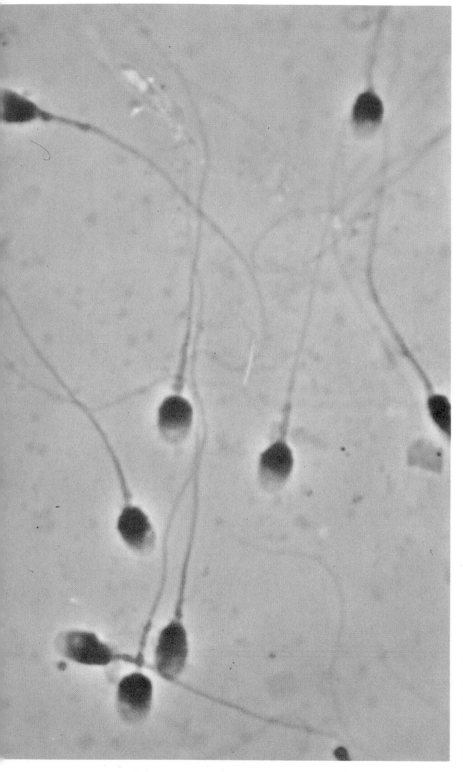

Portable instructions.

Opposite   *A mammalian egg.*
When the genetic constitution is
complete – when both *the*
delegates from the previous
assemblies have pooled their
instructions – the process of
reconstituting the next assembly
can begin.

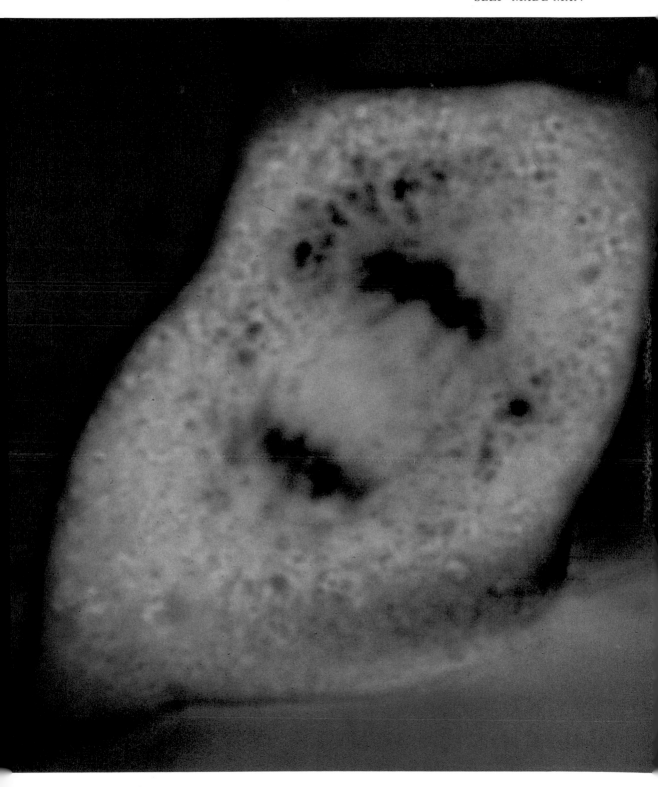

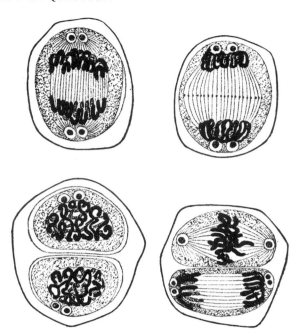

which enabled scientists to reject the pictorial model of heredity in favour of one based on linear programming. For example, the discovery that each and every cell was invariably furnished with a darkly staining nucleus and that this participated in every division led them to suspect that the directions were contained in this structure. The discovery of fertilisation had already cast doubts on the existence of a preformed image, since it was difficult to see how the individual unity of such an image could be reconciled with such a patently dual origin — unless one assumed that the effigies were brought together in two halves, in which case maternal characteristics would invariably be distributed down one side of the body with the father's features on the other. In any case, microscopic observation had already shown that, although the nucleus had a complicated structure, the material bore no resemblance to the adult shape but took the form of minute ribbons or chromosomes. From the way in which these entities reduplicated themselves and were accurately shared out each time the cell divided, scientists were unable to escape the conclusion that these were the sites of the genetic inscription. And although no one had yet discerned, let alone deciphered, the characters, it was almost self-evident that the instructions were assembled in a factorial rather than a pictorial manner.

If scientists had heeded the plant-breeding experiments of an obscure Moravian monk in the 1860s, the factorial view might have established itself forty years earlier than it did. Until Gregor Mendel, almost no one had recognised that bodily characteristics could be inherited and distributed amongst the offspring in a strictly arithmetical manner. By carefully choosing characteristics which displayed themselves in a discrete all-or-none manner, and by working with breeding populations which were large enough to display convincing arithmetical trends, Mendel was able to prove conclusively that the hereditary material — of which he had no immediate knowledge — was composed of genetic particles which could be transmitted independently of one another. Such experiments, with their unprecedented commitment to the statistical method, would have dealt a death blow not only to preformationism but to the whole controversy which had dogged embryology since Harvey and before. By the end of the nineteenth century, however, the discovery of chromosomes all but compelled the rediscovery of Mendel. And by 1900 the stage was set for a new genetic theory.

What eventually emerged was not simply a new theory but what the American philosopher of science T. S. Kuhn has called a new paradigm. That is to say, not a hypothesis about

Mendel's classic experiments with peas showed that the hereditary instructions are broken up into discrete units and that the programme is compiled in an arithmetical fashion rather than a pictorial one.

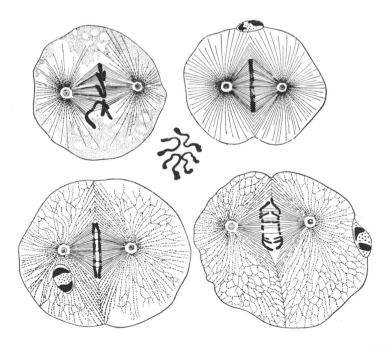

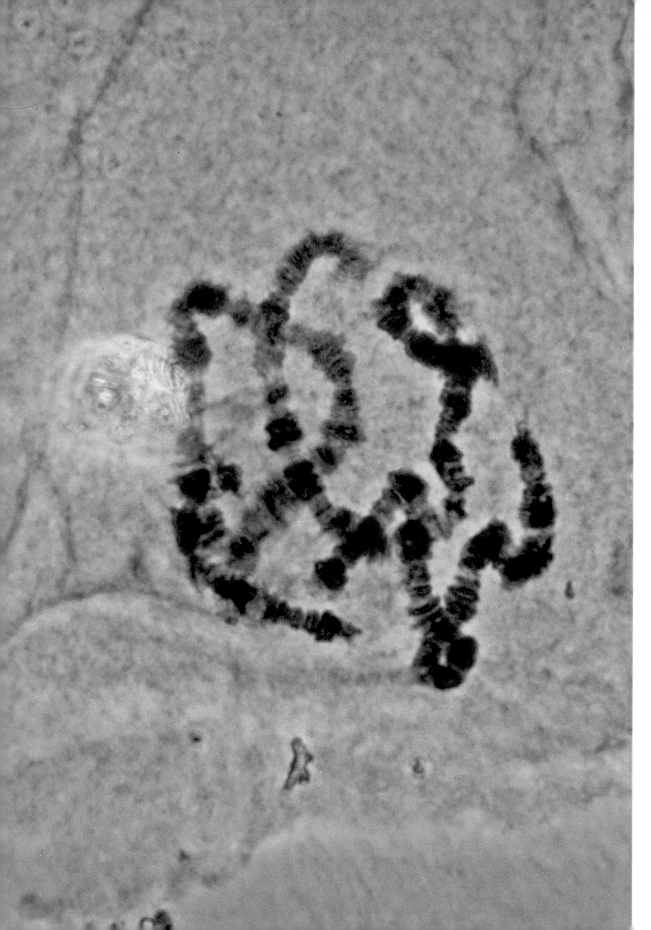

At the end of the nineteenth century microscopists discovered that the nucleus of every living cell contained ribbon-like bodies with a special affinity for certain chemical dyes – hence the name chromosomes. *The fact that these objects were reduplicated and then carefully shared out at each cell division led to the recognition that they* carried the hereditary information.

the way things work, but a shift in mental attitude which admitted hypotheses of a completely different order. For instance, without any observations to justify the assumption at that date, certain late nineteenth-century embryologists were already beginning to speak of the instructions in alphabetical terms, suggesting that, whatever material form the code assumed, the characters which composed it were a finite set but could be combined to form an infinite variety of instructions.

In one sense, of course, the material or substance out of which the characters of any notation are made makes no difference to the communicative power of the code. As long as the individual characters are unmistakably distinct from one another and as long as there is an agreed significance attached to the occurrence of any one of them, what they are made of is irrelevant to the message which they help to carry. *Hamlet* remains the same whether the text is chiselled in stone or scribbled on parchment. A satisfactory performance can be reconstituted in accordance with any of these inscriptions.

Interestingly, the recognition of the formal or conventional nature of coded information was being made in other fields at the very moment when geneticists were beginning to see its relevance to the study of heredity. In 1913 the Swiss linguist Ferdinand de Saussure pointed out that the characters which compose a code bear an arbitrary — or, as he said, unmotivated — relationship to whatever they signify or represent. De Saussure hastened to point out, however, that although the relationship between a linguistic sign and the concept which it signified was an arbitrary one — so that there was no way in which an English speaker could tell simply by looking at the word '*soeur*' that it meant 'sister' — the relationship was not haphazard; within the French-speaking community there was no getting away from the fact that '*soeur*' referred to female sibling. So that, although geneticists were beginning to suspect that the linguistics of heredity were arbitrarily coded, it was nevertheless a matter of urgency to discover in what characters the inscriptions were written.

The story of DNA has been repeated so often that there is no need to reiterate the heroic details of the discovery. Unfortunately, its reputation as a double helix has obscured one of its most significant features. People have got the impression that the mysterious potency of this macro-molecule is associated with its distinctive shape, and that the genetic instructions are somehow conveyed by the seductive curves of the spiral itself. But the fact that the genetic instructions are carried on a spiral is irrelevant to the part which they play in heredity. It makes no difference to the readers of the Bible whether the text is unreeled on a scroll or bound leaf after leaf in a book: what matters is the distinctive legibility of the characters — and the most significant part of Crick and Watson's achievement was the identification of the molecular notation. Without minimising the biophysical importance of the helical arrangement, it is essential to point out that their revolutionary contribution was the recognition that the genetic inscription had only four characters, but that by grouping these in serial combinations it was possible to encode all the information needed to guide both development and adult function.

The chromosomes which had been identified at the end of the nineteenth century were now seen as a twisted pair of ribbons, serially linked to each other throughout their length by duets of complementary molecules known as nucleotides. From the orderliness of this sequence it became apparent that

Opposite   Messenger RNA carries a transcript of the nuclear instructions into the body of the cell. At special sites in the protoplasm the messenger RNA sets up a synthetic centre while amino-acids are assembled in orderly sequences to produce polypeptide chains from which proteins will develop.

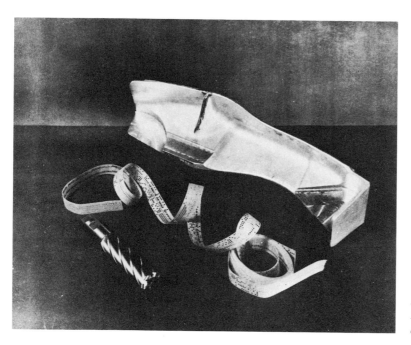

An industrial 'chromosome'. Machine aircraft-forging with cutting tool and 'punch' tape.

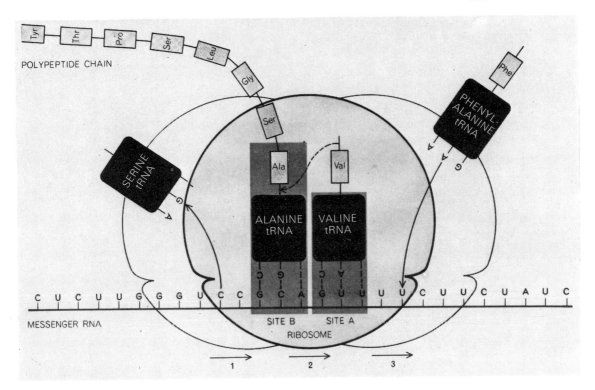

ALANINE	Ala
ARGININE	Arg
ASPARAGINE	AspN
ASPARTIC ACID	Asp
CYSTEINE	Cys
GLUTAMIC ACID	Glu
GLUTAMINE	GluN
GLYCINE	Gly
HISTIDINE	His
ISOLEUCINE	Ileu
LEUCINE	Leu
LYSINE	Lys
METHIONINE	Met
PHENYLALANINE	Phe
PROLINE	Pro
SERINE	Ser
THREONINE	Thr
TRYPTOPHAN	Tryp
TYROSINE	Tyr
VALINE	Val

it carried a message, and in the last twenty years the structural grammar of the communiqué has been fully elucidated.

As in spoken or written language, the smallest combination of letters capable of carrying a meaning is a word. Molecular biologists call this irreducible unit of genetic meaning 'the codon'. The entire vocabulary is made up of three-letter words which ring the changes on all the possible combinations of the four available nucleotides. Each of these codons corresponds to or 'means' one of the twenty so-called amino acids which form the constituent elements of protein. But isolated words do not comprise an instruction: there are no exclamations in genetics. To issue an order it is necessary to utter a well-formed grammatical sentence, long or short, which can't be abbreviated without destroying its meaning. Each one of these irreducible sentences constitutes the instruction for the manufacture of a recognisable protein. The readable separation of these sentences is guaranteed by a punctuation which is also built into the code. Some of the codons are introduced into the sequence for the sole purpose of marking the end of one sentence and the start of another. They convey no information in their own right, but ensure that there is no confusion between the messages which they separate.

At a still higher level of organisation the chemical sentences are grouped into distinct paragraphs, each one of which corresponds to a recognisable bodily characteristic. With the rediscovery of Mendel's breeding experiments, geneticists inferred the existence of such paragraphs, without knowing anything about their material constitution. They noticed that the factors which made up the demeanour of an individual were often segregated from one another in the offspring of the next generation, indicating that there was some discrete vehicle in the genetic material which could be transmitted independently of all the others. These hypothetical entities were known as genes, and until the discovery of the DNA code the way in which they conveyed their instructions remained obscure.

The linguistic metaphor is useful as far as it goes, but introducing terms like 'uttering' or 'reading' can give the misleading impression that a cognitive process is involved in following the instructions — and this would reintroduce the psychic agencies of Aristotle and Harvey. In the living cell the performance does not take place by recognising what is written, but by transcribing the message into a comparable material called messenger RNA, which is then despatched into the metabolic material of the cell, where its orderly series of codons dictates the regular synthesis of proteins.

Within such an arrangement it becomes very hard to make a distinction between structures which convey information and the ones which come into being by obeying it. Through the medium of a biochemical transcription service, one order of information is transformed into another, but since the material from which DNA is made has a different biochemical character from that of the proteins whose synthesis it dictates, there is no resemblance between the two. The connection between them is a matter of logical transformation. Although the two structures do not resemble each other, there is a strict rule of correspondence between their respective structures. As the mathematicians would say, the two entities are 'mappable' one on to the other. For that reason it is more profitable to think of a system as a computer program in which a series of

Right   Using a combination of
chemistry and crystallography
Crick and Watson in the 1950s
proved that DNA consisted of
two long strands of polymerised
sugar spirally locked to one
another by a staircase of
molecular 'treads'. Each
'tread' is spanned by a
complementary pair of
biochemical 'bases', of which
there are four recognised types.
Base A will only form a tread
if it is paired with Base T.
Base G is always found in
partnership with Base C, etc.
Prior to cell division the
staircase is split down the
middle of the flight of treads.
The double helix is then
reconstituted by attacking
appropriate fragments to the
broken 'treads'. The system
thus ensures that the original
sequence of 'coded' duets is
preserved.

Below   A model of DNA.
The famous image of the double
helix represents a triumph of
imaginative mathematics. With
this discovery biology became a
protectorate territory of
chemistry and physics.

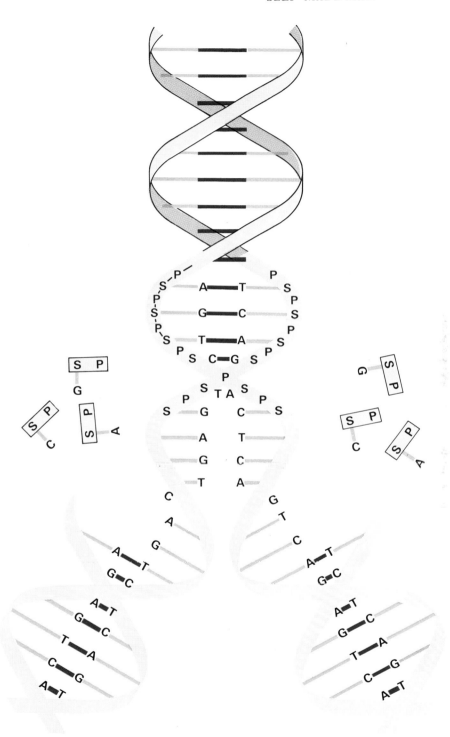

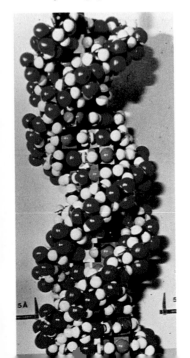

linear routines imposes order upon a material which would otherwise tend towards structureless incoherence. But in a universe governed by the Second Law of Thermodynamics, the imposition of structure and order can be achieved only with the expenditure of energy, and although the DNA prescribes *which* order is to prevail, it can bring it into existence only if energy is spent on the enterprise. The pedantic precision of DNA can become effective only in the presence of the energy provided by ATP.

The traffic in information does not flow in one direction, however. The whole system is regulated by feedback loops which ensure a dynamic equilibrium between the nucleus and the protoplasm. The principle of homeostasis repeats itself at the smallest conceivable level of biological organisation. Through the medium of these self-regulating exchanges the linear order represented in the nucleus re-expresses itself in the three-dimensional order of the protoplasm. Through the mechanism of successive divisions the cell transforms this level of order into the federal structures of the corporate organism.

Because the outcome is achieved through the orderly succession of routines and sub-routines it is no longer necessary to look for signs of the adult form in the early stages of its own development. The best way to see what happens is to repeat the simple experiments done by the late nineteenth-century embryologists. Because its early cell divisions are so easy to observe, the sea urchin became one of the experimental models for embryological research. If you crack open a sea-urchin and squeeze its microscopic eggs into a jar of sea water, you can fertilise them by milking the roe of a male urchin into the same jar. If you draw up a drop of the mixture and put it on a microscope slide, you can study the events hour by hour. All you can see to begin with is a crowd of ova suspended in the water like translucent planets, each one surrounded by a shimmering halo of spermatozoa. One of these penetrates the outer membrane, fuses with the nucleus of the egg cell, and fertilisation is complete. Within ten to fifteen minutes, the cells begin to divide into two equal halves. These divide in their turn, and so on — forming a hollow ball of little cells rather like a blackberry.

The ball begins to reshape itself: longitudinal grooves and

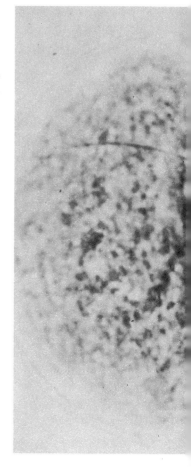

Opposite *Successive divisions of the fertilised egg produce the cellular units which will eventually comprise the living individual. Each unit contains identical instructions, but before long the cells will begin to differentiate into recognisable tissue types. It is as if the musicians in the orchestra had all been issued with full copies of the score, but know intuitively which stave to play.*

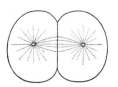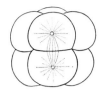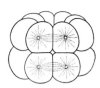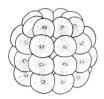

The re-shuffling of genetic material which accompanies sexual reproduction means that no child is an exact copy of its parents. But if the fertilised egg falls apart after the first cell division, the resulting children will be made by identical instructions and will resemble one another more than they do their parents.

symmetrical dimples begin to appear; pleats, seams and scrolls organise themselves into the primordial traces of what will later become the adult organs. The linear order of the sub-microscopic DNA has already yielded the three-dimensional order of a visible organism. Reaching down through successive levels of organisation, a vertical line of logical transformation connects the one to the other. The mystery is that each cell carries identical copies of the instructions which were contained in the undivided original, and yet by the

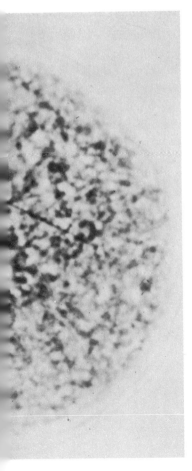

time they have divided into a congregation of sixteen — and in some species much earlier — the cells have begun to confine their attention to particular sub-routines of the total program. Henceforth, each cell is bound to differentiate itself into the tissue type for which it is destined and, as long as the program is correctly written, accurately transcribed and slavishly followed, the adult creature emerges in the image of its parents.

But the instructions contained in the sperm or ovum may be misprinted — a so-called mutation. Since the genetic message has a considerable measure of redundancy, a certain number of typographical errors can be tolerated without loss of intelligibility. It is easy to make sense of a long-winded letter which is filled with spelling mistakes, but when the redundancy is eliminated, as it is in a telegram, small slips can lead to major misunderstandings. Even in a richly redundant communiqué, there are limits to the number of misprints that can be tolerated. If the error results in a completely nonsensical sentence, the whole program may be aborted at the outset. However, the mutation may yield a sentence which reads quite sensibly, but which if acted upon brings about a recognisable error in the form or function of the developing organism.

Misprints on this scale usually produce such devastating results that the creature dies before it can breed and perpetuate the error. But if an organism bearing such a misprint reaches sexual maturity it will bequeath the error to its offspring. Illnesses like haemophilia and Huntington's Chorea are among the best-known examples. A rare mutation in the instructions regarding the manufacture of the red-blood pigment leads to a complicated condition known as

Twinning is only acceptable when the separation produces independent copies.

```
The terrible news is that mother fell and broke her leg at the
hotel in Tunisia and the doctors say she cannot be moved home.

The terribul new is that mothen fell and brole her leg at the
hoten in Tunezia and the docdors sae she cannor be mored home.

She tettibre newg if shat bother felt had krobe ehr gle at teh
hot elintu nosia and eth codorst say the callot eb vomed hoem
```

```
   MOTHER FELL BROKE LEG TUNISIA HOTEL CANNOT MOVE HOME
   MOTHER BROKE TUNISIA HOTEL FULL CANNOT MOVE COME BEG
   NOTHER EFLL BROKA BEG TUMINIA HOT CARROT MRVE
```

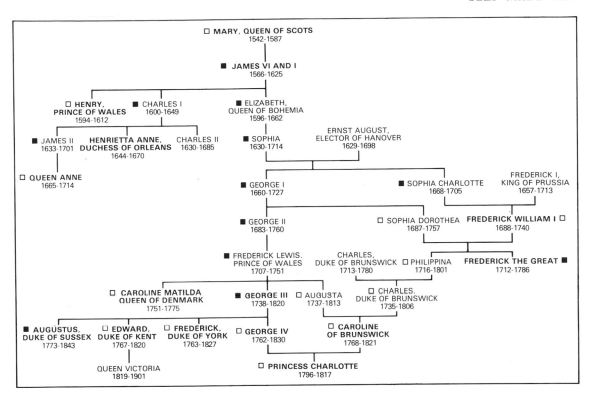

The mutation responsible for the disorder known as porphyria entered the English line from the Stuarts and then spread in a genetic cascade through the following generations.

porphyria, one of whose symptoms is episodes of lunacy. This error is particularly notorious since it entered the genetic stream of the English royal family from the House of Stuart. After the exile of James II it might have dispersed itself amongst the Continental royalty if the authorities had not been so pedantic in their search for a dynastically legitimate successor. By insisting that the new line should have connections with the Stuart ancestry, porphyria was re-seated on the English throne, culminating in the exorbitant madness of George III.

Very few congenital abnormalities are the result of inherited misprints, however. In most cases, a faultless program is inherited from the previous generation, but it may be carelessly transcribed or inaccurately obeyed during the course of development. The origin of such faults is usually unknown, but the cause can sometimes be identified in drugs, radiation or infection — for instance, the unsuspecting use of thalidomide, unnecessary exposure to X-rays during pregnancy, or German measles. If the error is incompatible with further development, the conception is aborted at a very early stage: a large number of unsuspected miscarriages — pregnancies which get no further than a slightly delayed period — represent inevitable drop-outs of this sort. But if the development is not stifled altogether, if the mistake or blemish is biologically tolerable, the infant makes its appearance with a congenital abnormality. The damaged foetus may have just enough efficiency to live for a few days after it has been delivered, or it may carry its fault into adult life.

In some cases, the defect remains acceptable until the changes of life which coincide with the normal milestones impose an intolerable strain: an infant born with a congenital abnormality of the hip, for instance, may remain free of symptoms until he learns to walk; a hole in the heart may not become symptomatic until the infant starts to run.

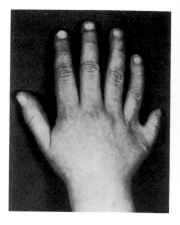

Hereditary polydactylly.

Such errors can affect anything from architecture to behaviour. If the routines are duplicated, inverted or incoherent, parts or organs may be doubled, reversed or dishevelled. There may be extra fingers, an additional lobe to the kidney, a womb divided in two; the normal asymmetry of the organs may be transposed, so that the heart and great vessels lie over to the right, and the appendix is on the left; there may be local blemishes in the architecture of the blood vessels — birthmarks and port-wine stains.

Some of the most noticeable errors, however, represent sins of omission: they are the result of leaving out or skipping essential passages in the developmental instructions. These result in local arrests of development, and for that reason provide interesting evidence about the way in which the foetus arrives at its finished form. Each part of the body has its own characteristic repertoire of such arrests. Errors in facial architecture invariably take the form of clearly defined gaps or interruptions in an otherwise normal human likeness, and since these clefts always appear in one of three or four sites and nowhere else, one can tell that these are the natural seams.

The human face, it appears, is the product of millinery rather than sculpture: it is not moulded from a single block of rough material, but pleated and stitched out of separate components. At a very early stage in the development of the head, sheets of skin and muscle start to organise themselves around the landmarks of the mouth, nose and eyes. A single visor grows symmetrically down the middle to form the forehead,

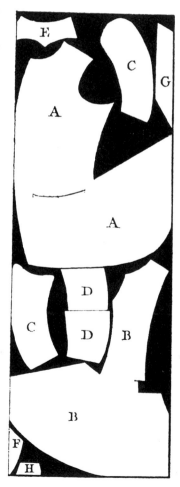

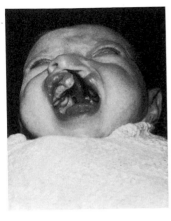

Unfinished sculpture provides useful information about the way in which the artist works. Congenital abnormalities show that human beings are made by a different sort of process altogether.
(Right   Michelangelo, Pietà Rondanini, Castello Sforzesco, Milan)

Opposite   Failure to complete a design like this would result in gaping seams. Human beings are not cut out of separate swatches of material – but the pleating and seaming of the cell layers means that the process is much closer to tailoring than it is to sculpture – biological seams gape too. (Pattern from F. A. de Garsault, 'Description des arts et métiers, l'art du tailleur', Paris, 1769)

the upper eyelid, the front of the nose, the little groove that runs down just under the nose, and the web that links the inside of the upper lip and the gum. At the same time, a pair of horizontal promontories grow in from the side under the eyes to meet in the centre, forming the cheek, palate and outer wings of the upper lip. If these processes fail to synchronise, if the anatomical rendezvous is frustrated, a harelip may appear either on the right or on the left; or there may be a large gap down the middle and the shelf separating the nose and mouth may fail to close, forming a cleft palate; occasionally, the seam may gape along its full length from the upper lip to the inner corner of the eye.

Aberrations appear with the same consistency in the development of the central nervous system. Once again, they are predictable failures to follow a preordained plan to its appointed conclusion. All of them, therefore, represent an earlier stage of development. In the young embryo the brain and spinal cord make their appearance as a shallow groove, excavated down the midline of the back surface. The edges of this groove roll over and grow together, so that the whole of the cerebral spinal axis assumes the form of a tube, which gradually zips itself shut from head to tail, leaving a microscopic pore at the lower end, which finally seals itself off. If the zipping instructions are not followed, the spinal cord fails to close, and the nervous system is left exposed at the lower end, forming the well-known condition of spina bifida. In this affliction the nervous elements fail to function in the affected area, and the lower limbs are paralysed; nervous control of the bladder and rectum may never develop properly, so that the child is incontinent.

The regularity of these errors shows that congenital abnormalities are not outlandish whims of the providential imagination, but failures in the execution of an orderly plan, intelligible only in terms of the biological statute or constitution from which they depart. Although the congenital abnormalities I have just described tend to recur with monotonous regularity, their appearance in one generation has no bearing on what will happen in the next: since the errors which are responsible for them are not written into the genetic instructions, they vanish with the death of the individual who bears them.

Mutation is a different matter altogether. When a mistake crops up in the DNA code it is transcribed not only within the cell that bears it, but replicated with each successive cell-division and handed on to the germ cells to be transmitted indelibly to the next generation. The source of such errors remains obscure, although there are well-recognised mutagenic agents, of which radiation is the most conspicuous example. However they arise, almost all mutations tend to have an adverse effect, and in almost every case the individual who bears them is extinguished before being given the opportunity of transmitting the defect to his successors.

Nevertheless, such unpredictable novelties are the only possible source of biological improvement. For every 100 departures from the plan that prove to be unworkable or regrettable, there is always a chance that one may be useful, and if the favourable novelty is repeated it helps to turn the race in a more profitable direction. This is what evolution is about. It is not the outcome of a conscious urge towards self-improvement, any more than individual development is the thoughtful fulfilment of a creative imagination. Novelties are introduced without any regard to their biological usefulness and establish themselves only if they confer a recognisable advantage. Purpose, it seems, is the cumulative outcome of chance.

The random acquisition of such advantages is painfully slow. Mutations which eventually turn out to be useful usually appear in forms which are scarcely perceptible, and gain a recognisable foothold in the genetic material of the race only after repeated recurrence and subtle recombination. The homeostatic mechanism of the genetic apparatus itself is intolerant of major upheavals and, unless its own balance is destroyed by the novelty, it exerts a restraining influence upon it so that, in the first instance at least, the new feature achieves the most modest expression in the life of the individual who bears it. But although the genetic homeostasis of the living cell tends to restrain the expression of profitable mistakes, when these *are* allowed to express themselves they generally do so by conferring ever greater homeostatic abilities upon the lucky lines that inherit them. For, in the long run, what makes a novelty acceptable, what wins it favour in the competition for existence, is the fact that it confers upon the organism an even greater ability to withstand the random destructiveness of the inanimate universe.

Of all the homeostatic endowments that have been thus acquired the most profitable is the one which enables the creature to build a cognitive model of the world that so inexorably threatens its existence. With the evolution of the human nervous system, nature accidentally created an apparatus capable of representing her own likeness — something which was capable of recognising and formulating the thermodynamic conditions in which it was forced to live.

THE MECHANISMS I HAVE DISCUSSED SO FAR ALL HELP
to create the physiological stability without which Claude
Bernard said that it was impossible for any creature to enjoy
a free life. What he did not point out is that the actions
which constitute such a free life are themselves undertaken
to ensure the physiological constancy of the internal environment. The ambition, the versatility and intelligence with
which the creature pursues food and avoids predators are
merely homeostatic aids to the maintenance of the *status quo*.
All action is directed towards a state of affairs where there is
no need for further action. For living things, all restlessness is
directed towards the achievement of tranquillity. Breathing,
healing, sweating and flushing; eating, drinking and urinat-
ing, are all undertaken in the service of an almost unattainable
stability. In the previous chapter we saw how the metaphor
of human action delayed and frustrated the recognition of
processes which were involved in biological development.
Nowhere has this metaphor been so obstructive as in the
discussion of human action itself.

If you were to ask a qualified driver what he had to do to
start his car, he would probably list a series of actions which
would include turning on the ignition, pressing the clutch,
and shifting the stick into first gear. For each of these steps it
would be possible to give an answer to the question 'How is it
performed?' To turn on the ignition, you have to twist the key
clockwise; to disengage the clutch you have to press down on
the pedal with your right foot. As you push the enquiry
backwards, however, there comes a point when the questions
no longer make sense. It sounds nonsensical to ask someone
what he has to do to twist the ignition key, or to press down
with his right foot. He would be able to demonstrate each of
these actions simply by doing them, but he would be unable to
describe the steps he had to take in order to get them done.

There are, it seems, two sorts of human action: those
involving recognised preliminaries, and those for which there
are no preliminaries. Driving cars, throwing stones, lifting
weights and emptying buckets are all examples of the first
sort, since in each case there are describable actions which
have to be taken in order to get them done. The other sort get
done by being done. Philosophers call them 'basic actions',
which means that they are irreducible, at least as far as the

*A panorama of mediated
actions. In order to complete
these tasks the performer has to
make muscular movements —
but he has not got to* do
*anything in order to make his
muscles move. He just moves
them. Something else may
happen in moving the muscles,
but he* does not make them
happen.

288

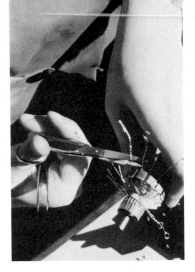

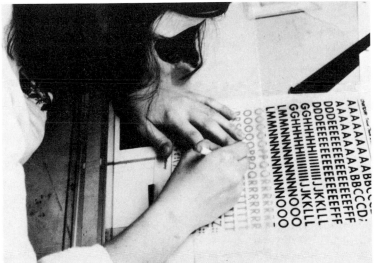

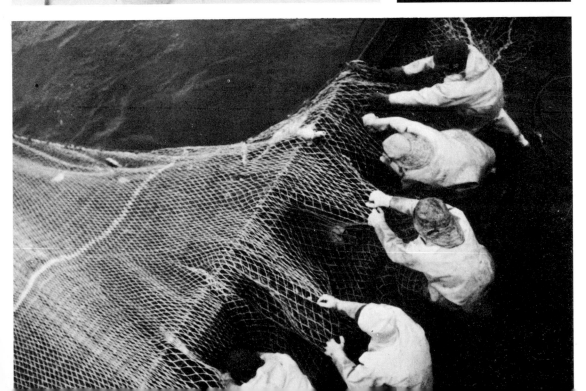

actor himself is concerned. Scientists can show that they are preceded by all sorts of physiological events — the contraction of muscles and the stimulation of nerves — but when someone moves his foot he doesn't choose to move his muscles first and then find to his delight that his foot has moved. What he *does*, or rather what *he* does, is to move his foot.

All of our effectiveness — everything we do, everything we get done — can eventually be traced back to something we do without consciously having to do anything else first. All our performances, it seems, are the spontaneous expressions of our wish to get them done. As far as our muscles are concerned, the causal effectiveness of our will seems to be so immediate that it must have taken an enormous leap of the imagination for men to stand outside themselves and recognise the possibility that their actions might be preceded by physical events of which they had no knowledge, events over which they as agents had no voluntary control. Far from trying to explain actions in terms of physical processes, human beings were understandably tempted to explain all physical processes as if they were motivated actions, the result of psychological urges. Even someone as sophisticated as Aristotle attributed the movement of physical objects, like the falling of a stone, to some inherent urge to arrive at the proper destination.

For primitive man, the universe was a province of motives rather than causes: man projected the image of his own effectiveness on to the universe around him, and, seeing his own behaviour as the pure expression of will, he assumed that the inanimate world operated in the same way — that it was not inanimate at all, but, like him, moved by a conscious purpose. At the same time, he had a growing suspicion that although will was a necessary condition of change, it might not be a sufficient one: even if soul was the universal prime-mover, there had to be some medium or instrument capable of translating its mental actions into physical events; some intermediate substance half-way between mind and matter, as weightless and transparent as thought but with just enough substance to move a heavy object or limb.

There have been many candidates for such a medium in the history of European thought. Some philosophers associated it with flame; others assumed that it was indistinguishable

from air, breath or *pneuma* — in fact until the middle of the nineteenth century psychology was often known as 'pneumatology'. Aristotle insisted that it was too insubstantial to be associated with any of the four physical elements, and that there had to be a fifth, a 'quintessence', a radiant, impalpable potency which was undetectable apart from the changes which it brought about.

The idea of such a medium has had a long-lasting influence on European thought, extending from Plato to the present day. Whenever philosophers have tried to re-establish the sovereignty of the soul, the concept of a universal 'ether' has invariably reappeared in one of its many guises. For the Neo-Platonists of the fourth and fifth centuries AD, it was a radiant emanation; for the Florentine mystics of the fifteenth century, it was 'a very subtle body; as it were not body and almost soul, or again, not soul and almost body ... it vivifies everything everywhere and is the immediate cause of all generation and motion'. For the intellectuals who gathered round Lady Conway it was, in Henry More's words, 'a substance incorporeal, but without sense and animadversion, pervading the whole matter of the universe and exercising a plastic power therein';

291

for Mesmer, it was the magnetic fluid, filling the wide vessel of the universe. In our own century it reappears as Wilhelm Reich's orgone.

This hybrid notion, claiming to reconcile the separate realms of mind and matter, contains the seeds of its own destruction. Mental actions and physical events belong to different universes of discourse, and there is no intelligible way of making a causal transition from one to the other. The introduction of a paraphysical medium which surreptitiously changes its character is simply a philosophical sleight-of-hand which subtly alters the terms of the argument by imperceptible steps. The interval between mind and matter can't be overcome by sneakily substituting one for the other. It would be just as illogical to suggest that there was an intermediate state half-way between numbers and apples: some substance which was simultaneously numerical and fruity — not quite 5, but not altogether Golden Delicious either. Whereas it is unlikely, but not logically impossible, that there should be a substance half-way between an apple and cider. To that extent, the concept of a spiritual ether is worse than wrong: it is nonsense.

Once such a concept was applied to the human body, however, it was described in a way that at least made it possible to be wrong — although what was wrong did not become apparent for nearly 1,500 years, mainly because the descriptions were too noncommittal to invite experimental criticism.

The first attempt to include this substance in a physiological system was made by Galen, who suggested that it was synthesised from food by distilling it until it was thin enough to be indistinguishable from mind but just substantial enough to move muscles. The process went in three stages, rather like the preparation of alcoholic liquor: in fact, the successive products were actually called 'Spirits'. As we have seen, Galen believed that food was concocted into blood in the liver, acquiring in the process an ingredient which he referred to as 'Natural Spirits'; when this material came into contact with respired air in the heart it was further enriched with 'Vital Spirits', and when this in turn was distributed to the brain it was finally converted into 'Animal Spirits' and stored in the cerebral ventricles. By the second century AD the brain had

Renaissance philosophers and magicians regarded the aether as a fifth mediating element which linked mind and matter. Within the body, where it took the form of Animal Spirits, it translated the impulses of the soul into movements of the muscles. In the universe at large it transformed the motives of God into the events of His material creation.

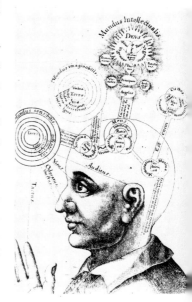

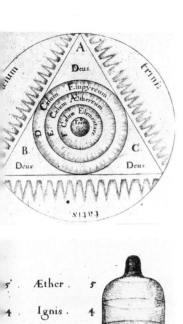

already been identified as the seat of the soul, and, since Galen had experimentally demonstrated the loss of both power and sensation when the spinal cord was severed, he concluded that the soul transmitted its orders to the muscles in the form of a pulse or shock-wave which travelled in the Animal Spirits all the way from the cerebral ventricles along the nerves into the muscles.

The first part of this scheme — the manufacture of Animal Spirits — exemplifies the logical absurdity that I have already mentioned: the interval between mind and matter can't be bridged by an industrial process. Furthermore, there is no way in which the theory can be shown to be either right or wrong, since Galen mentions no distinctive properties that would enable a scientist to tell whether any of the three Spirits were present or not. How would one distinguish blood containing Natural Spirits from pure blood? And although the cerebral ventricles are undoubtedly filled with fluid, Galen doesn't tell us how we could distinguish Animal Spirits from any other liquid.

Up to this point, therefore, the theory is vacuous, since it doesn't significantly assert anything but simply re-names the three physical substances which are to be found in the liver, heart and brain respectively. In the second half of the theory, however, Galen commits himself to the possibility of being wrong, since the scheme he proposes is so recognisably hydraulic in character that it implies there are certain physical features whose presence can be tested, and certain physical events which can be observed. Galen performed what he thought were the relevant tests, and the outcome of these convinced him that his original assumption was correct. He opened the neck of a live pig, and tied a ligature around the nerve which connected the brain to the muscles of the voice-box. The pig's squeals ceased at once, and since they resumed as soon as the ligature was loosened Galen assumed that he had blocked the hydraulic transmission from the cerebral ventricles to the larynx.

As he saw it, the experiment corroborated his theory, which only goes to show that theories do not stand or fall by the results of so-called 'crucial experiments'. Obviously, the theory would have been seriously jeopardised if the pig's squeals had survived the ligature, but the fact that they were

293

extinguished by it is open to several interpretations: the same manoeuvre would have blocked almost any physical transmission — electrical, thermal, chemical or hydraulic. The point is that in Galen's time hydraulic transmission was the only intelligible way of conveying a physical influence along such a narrow passage, so the experiment was bound to confirm his theory, if only for want of an alternative one.

For this reason the doctrine of Animal Spirits survived more or less unquestioned until the middle of the seventeenth century. When Vesalius described the structure of the brain he continued to regard it as a hydraulic cistern, and 100 years later, in his *Treatise on Man*, Descartes popularised the same model with a vivid metaphor:

> You may have seen in the grottoes and fountains which are in our royal gardens that the simple force with which the water moves in issuing from its source is sufficient to put into motion various machines and even to set various instruments playing or to make them pronounce words according to the varied disposition of the tubes which convey the water. And, indeed, one may very well compare the nerves of the machine which I am describing with the tubes of the machines of these fountains, the muscles and tendons of the machine with the other various engines and springs which serve to move these machines, and the Animal Spirits, the source of which is the heart and of which the ventricles [of the brain] are the reservoirs, with the water which puts them in motion.

Descartes went much farther than his predecessors and added a number of important physical details. For one thing, he located the soul in the pineal gland of the brain, insisting that this immaterial agency had the ability to open and shut the pores or entrances leading from the cerebral ventricles into the hollow nerves. Descartes supposed that by playing on this console of perforations the mind could compose distinctive melodies of muscular action, distributing Animal Spirits hither and thither depending on what movement was required. The logical absurdity of the original scheme was preserved — how could ghostly fingers operate physical valves? — but because it was a matter of logic no experiment could have a bearing on its truth or falsehood. Descartes did,

The civil engineers of Imperial Rome built a system of urban water supply which was not equalled until the seventeenth century. The image of reservoirs and aquaducts may have played an important part in shaping Galen's theory of nervous action. It certainly dominated the neurological thought of René Descartes.

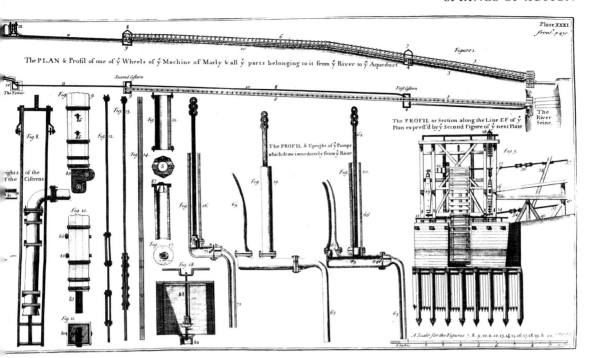

however, commit himself to certain physical assertions which were susceptible to experimental proof. In his scheme, the muscles were actually puffed up by the inrush of Animal Spirits, and this at least could be checked by physical observation. But unlike his Puritan colleagues on the opposite side of the Channel, Descartes shunned experiment with Jesuitical disdain, preferring to arrive at his conclusions by deducing them, like Euclid, from fundamental axioms which had in their turn been born from pure contemplation.

Other scientists were less fastidious. The seventeenth-century Dutch physiologist Swammerdamm stimulated the cut end of a nerve, and found that the muscle contracted without any increase in its overall volume; moreover, it continued to contract long after the nerve had been disconnected from the reservoir of Animal Spirits which was supposed to exist within the brain. This ruled out a hydraulic transmission, and showed that neuro-muscular activity depended on a different principle altogether. What could this principle be? As chemistry and physics stood in the middle of the seventeenth century, it was impossible to suggest a detailed alternative. The only known way of propagating physical

295

A. *Aries Simplex*. B. *Aries Compositus*.

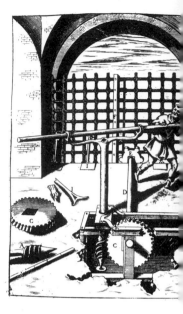

Until the discovery of electro-chemical energy, action at a distance was only intelligible in terms of pushes and pulls, impacts, torques and tugs.

change was mechanical — tugging, pushing, nudging, puffing or blowing — and, since the experiments had ruled out hydraulic transmission, it was hard to visualise what it was that travelled along the nerve and into the muscle.

There were, however, vague intimations of what later became the essential feature of modern neuro-physiological theory: the notion of inherent biological irritability according to which the capacity of living matter to respond to a stimulus is built into the material from which it is made. The first to make an explicit declaration of this was the English physiologist Francis Glisson who, around 1640, insisted that what made muscles contract was not something puffed into them from outside — they were not pneumatic balloons — but some property inherent in their substance which made them shorten when stimulated by their nerves. This, however, only pushed the problem one stage further back. How did the nerve exert such an influence on the muscle to which it was attached?

In the eighteenth century, the Swiss physiologist Von Haller suggested that nerve, like muscle, was endowed with a *vis nervosa* — an intrinsic biological power which had the ability to fire or detonate the irritable substance of any muscle with which it made contact. Some scientists were understandably troubled by this suggestion because it offered no explanation of the physical processes which were involved. The introduction of a special term like 'irritability' doesn't explain muscular action, any more than the term 'locomotive power' accounts for the movement of a railway engine; and christening the sensitivity of nerve with a Latin name doesn't say what actually happens. Nevertheless, the doctrine of irritability shifted the emphasis from an explanatory model based on hydraulic shocks to one where the intrinsic physical properties of nerve and muscle played the most important part. The fact that eighteenth-century scientists were unable to make sense of this suggestion is beside the point. The history of science is punctuated by theories which represent an '*agnosticisme provisoire*': ideas which seem to have no theoretical force at the time when they are expressed, but which nevertheless leave a gap that can be filled in later with experimental details. The notion of intrinsic irritability not only marks a significant break with the hydraulic tradition, it anticipates, albeit vaguely, modern notions of cellular energy.

Eighteenth-century scientists found it hard to accept such a doctrine, since there was no known process in the physical world with which it could be compared or identified. There were, however, technological devices which could have provided an alternative picture. William Harvey had even referred to one when he described the sequence of events in the heart-beat. He compared what happened to the firing of a pistol.

If this metaphor had been followed to its logical conclusion, it would have provided a fruitful model of what happened when a nerve stimulated a muscle. Unfortunately, Harvey used it only to emphasise the rapidity with which the contraction of the ventricles followed the contraction of the auricles: he had no reason to suspect the underlying processes which were involved. We now know that what happens when a nerve stimulates a muscle is, indeed, very similar to what happens when a firing-pin detonates a charge of gunpowder,

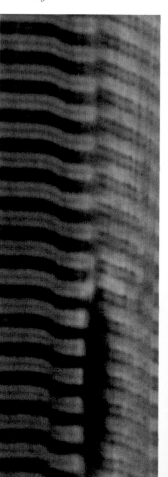

Muscle 'end-plates'. These specialised nerve endings are the sparking plugs or detonators of the muscular charge. The nerve impulse travels as far as the end-plate and then stops. On arrival it causes the release of a transmitter substance which travels rapidly across the gap — the neuro-muscular junction — and detonates the stored energy in the muscle. Each motor neurone is distributed to a fixed number of muscle fibres, forming a single motor unit, all of whose components 'fire' together — or not at all.

The characteristic stripes of voluntary muscle.

and that the similarity is not simply a question of comparable rapidity: the processes resemble each other in their use of chemical energy and, for that reason, there is an interesting mathematical relationship between the size of the stimulus and the magnitude of the response.

The force with which the ball is discharged from the pistol is much greater than the force with which the firing-pin strikes. The movement of the projectile is caused by the explosive charge and not by the mechanical impact of the hammer: as long as the impact is forceful enough to detonate the gunpowder, the gun will go off, and the power of the explosion is entirely proportional to the amount of explosive in the cartridge — it is an all-or-nothing affair. The same thing happens when a muscle goes off. An explosive charge of

The membrane surrounding a nerve fibre actively maintains a difference in chemical composition between the substances inside the fibre and the fluids bathing its outer surface. The interior has a high concentration of potassium (K^+) and a relatively low concentration of sodium (Na^+). In the external medium this situation is reversed. This reciprocity of chemical composition creates an electrical tension across the membrane — so-called 'resting' potential. If the permeability of the membrane is locally disturbed, allowing potassium to rush out of the nerve while sodium leaks in, the electrical potential collapses. If the 'collapse' of tension is intense enough it sets up a comparable va-et-vient in the adjacent segment of membrane, and so on. A wave of depolarisation sweeps along the cylindrical membrane, creating the nerve impulse. The impulse is pursued by a wave of repolarisation as the membrane repairs itself and restores the concentration difference upon which the resting potential depends.

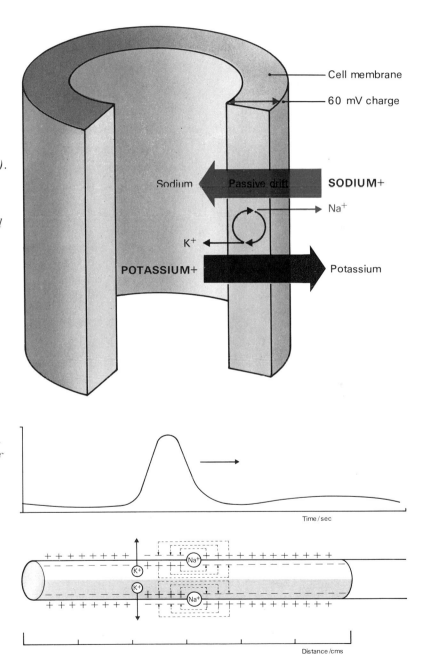

299

chemical energy is stored in its fibrous substance, and as long as this is provoked by a nervous stimulus of the right size, it will always release the same amount of mechanical work. The only way to increase the power of a muscle is to multiply the number of fibres involved — like pulling two triggers simultaneously on a double-barrelled shot-gun.

The pistol analogy breaks down at this point, however, because the nerve supplies no mechanical impact to the

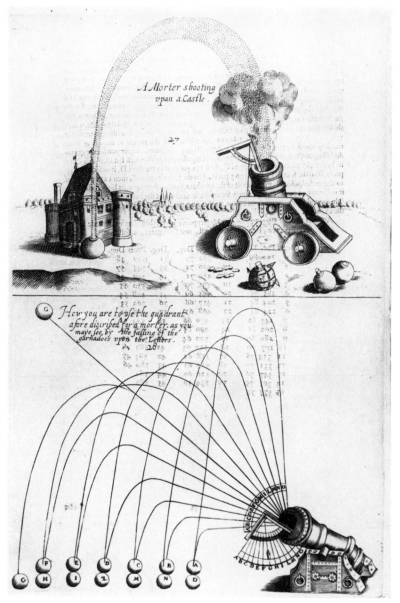

A primitive example of the 'all-or-none' principle in action. The range of a missile is determined by the size of the explosive charge and not by the strength of the impact which is applied to it. The energy for the explosion is stored in the chemical compounds which make up the gunpowder. This metaphor was available in the seventeenth century but since nerves and muscles did not visibly explode the relevance of the analogy was not seen for another 300 years.

muscle. The nerve doesn't move. It doesn't strike the muscle. It acts like a trail of gunpowder, and the impulse travels along it like a wave of heat. But, as in the case of muscle, the size of the wave is independent of the size of the stimulus which sets it going. Nerves, like muscles, obey the all-or-none principle. The intensity of the impulse depends on the amount of biological fuel distributed along the line. As long as the original stimulus is large enough to detonate the charge, the nerve responds to the utmost, with the impulse replenishing itself as it goes along by consuming the fuel laid out ahead. When the wave of biological incandescence arrives at the muscle, the muscle blows up like a keg of gunpowder.

The relevance of this metaphor seems immediately obvious to modern scientists, but it is easy to understand why it was overlooked by their seventeenth-century predecessors. Before an image can act as a metaphor, it is necessary to identify some process to which it bears an obvious resemblance. Looking at nerve and muscle with the naked eye, no one would think of comparing them with either a pistol or a gunpowder trail: the nerve doesn't change colour, and it doesn't make a sound; it doesn't glow, blacken or sizzle, and the contraction of muscle is not accompanied by a flash of light and a bang. The disturbance, such as it is, is too small and too subtle to have been detected by any of the instruments which were available in the seventeenth century, and, although modern instruments can detect the fact that there are slight changes of temperature accompanying the transmission of a nerve impulse, this is not the most important part of the process.

What travels down a nerve is a wave of electro-chemical activity, and although this has the same quantitative characteristics as a burning fuse the underlying processes are very different. Until the eighteenth century, scientists had no reason to suspect that living matter could behave electrically. Electricity had been associated with thunderstorms and frictional machines, and it therefore came as a great surprise when in the late eighteenth century the Italian scientist Galvani found that frogs' muscles could be stimulated by the application of electrical currents. It was not until the nineteenth century, however, that the electrical character of the nerve impulse itself was appreciated; and the all-or-none behaviour did not become apparent until scientists succeeded

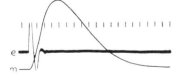

The basic electro-mechanical events of muscle action. The tracing 'e' shows the 'all-or-none' electrical spike of the arriving nerve impulse. The wave 'm' shows the rise and fall of mechanical tension as the muscle fibres shorten and then relax.

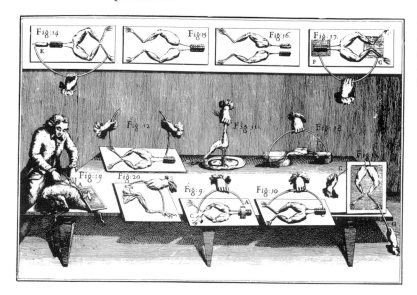

Galvani's concept of animal electricity proved to be infinitely more fruitful than Mesmer's notion of animal magnetism.

in creating instruments which could display and measure the size of the impulse.

Shortly after the First World War, the English neurologist E. D. Adrian showed that the electrical intensity of the nerve impulse depended on the size of the nerve rather than upon the strength of the stimulus. The stimulus simply released the wave of electrical disturbance and set it on its self-perpetuating path. Each nerve invariably displayed impulses of the same size and, since the disturbance replenished itself as it travelled, it remained the same size from one end of the nerve to the other — unlike a shock-wave, which would have slowly died away in transit. Adrian and his colleagues immediately saw that this simple physiological fact imposed an important constraint on the nervous system. If the size of the nerve impulse is always the same regardless of the stimulus which provoked it, there has to be an alternative way of registering intensity. Adrian soon discovered that it was a question of frequency: the stronger the stimulus was, the more impulses were despatched down the nerve, so that the intensity of the stimulus was conveyed by the number of impulses travelling, rather than by the size of any one of them.

The only way of strengthening a muscular reply, then, is either by recruiting extra fibres or by discharging stimuli into the muscle at such a rapid rate that the fibrous material has no time to relax in between. Each new contraction is, therefore,

Since muscle fibres obey the 'all-or-none' law there are only two ways of increasing the strength of contraction: either by recruiting more fibres or by firing such a rapid volley of nerve impulses down the motor nerve that the muscle fibres have no time to relax between 'shots'. Each contraction is thus added to the unrelaxed state of its predecessor, causing a tetanus.

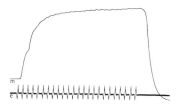

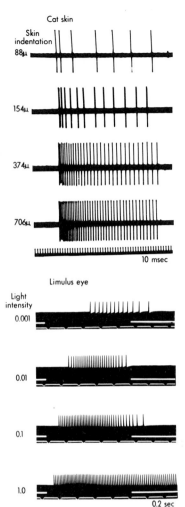

Cat skin

Skin indentation

88μ

154μ

374μ

706μ

10 msec

Limulus eye

Light intensity

0.001

0.01

0.1

1.0

0.2 sec

The 'all-or-none' law applies to sense organs too. The intensity of a stimulus is registered by the number of impulses which it generates and not by the size of any one of them. As the skin is more deeply indented the number of nerve impulses increases, although each impulse is the same size as its predecessor. As the light falling on the retina is brightened the impulses firing into the brain become more frequent.

superimposed on its predecessor, and the muscle builds up a so-called 'tetanus', which is simply another name for an unremitting state of contraction brought about by the rapid volley of motor impulses.

The all-or-none principle also applies to sensory nerves; that is to say, to all those fibres bringing information into the nervous system. When a spot of light brightens on the retina, the impulses travelling along the optic nerve multiply in number without increasing in size. When a joint is bent or a muscle is stretched, the intensity of the mechanical event is conveyed in numerical terms. In other words, both action and experience are counted out rather than weighed. Like the instructions which transmit the hereditary details from one generation to the next, the information moving within the nervous system is conveyed in coded bits. There would be no way of telling either the quality or the magnitude of the information which it is carrying simply by inspecting the bits. The information is conveyed by the overall sequence of what is transmitted — by the frequency of the runs, the intervals between them, and above all by the fibre which bears this traffic. For instance, the impulses which travel down the optic nerve have no detectable quality of luminescence, and the transmissions in the auditory nerve betray no qualitative evidence of their acoustic origin. The impulses in each are indistinguishable from one another. The subjective impression of light is the inevitable result of impulses arriving from the optic nerve. In fact, the system can very easily be cheated: if the retina is stimulated by mechanical pressure, say by pressing the eyeball with the tip of the finger, one immediately sees a patch of light.

The experiments I have just described show the sort of currency the nervous system uses to transact its business, but they do not explain how or why the impulses are distributed in the way they are. Descartes made a lasting contribution to this problem. Like his predecessors, he noticed that certain sensory stimuli tended to produce the same muscular response, and in order to explain this consistency he suggested that there must be a pre-established loop of nerve fibres linking the sensitive surface to the muscles which responded, and that this loop passed through the central nervous system. It had been recognised since antiquity that people could flinch,

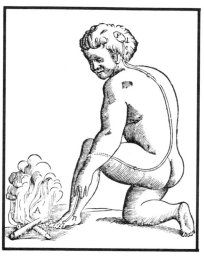

Illustrations from Descartes's Treatise of Man.

blink, scratch, sneeze and cough without having to think about it, and Descartes's model provides an admirably modern account of all such involuntary responses. But his theological scruples prevented him from regarding human beings as robots: he conceded that animals acted automatically but insisted that man was distinguished from these brute machines by owning a rational, immaterial soul located in the pineal gland. Descartes bequeathed science an awkward philosophical problem by making this distinction, but it nevertheless had certain advantages. By locking the conscious mind in a metaphysical penthouse, he made it possible to consider a large part of the human repertoire in a mechanical fashion. From then on, it was possible to describe automatic actions in terms of a neuro-muscular loop, and it set scientists looking for the arch of fibres linking the sense organs to their appropriate muscles.

At the beginning of the nineteenth century, two scientists working independently of each other in France and in England, François Magendie and Sir Charles Bell, showed that the arrangement was even more arc-like than Descartes had suspected. They found that sensory stimuli such as pain entered the central nervous system through channels reserved for them alone, and that motor impulses were segregated in the same manner. Although sensory and motor fibres are bundled together in the same cable while travelling in, say, the limb, they part company shortly before they reach their

Descartes had seen the necessity for some mechanism which allowed a muscle to shorten freely without having to overcome the obstinate resistance of its antagonist. For him the reciprocity was achieved *within* the muscles themselves, *by shunting animal spirits across a complicated 'clover leaf' of aquaducts. Sherrington showed that the give-and-take occurred within the spinal cord — by inhibiting the nervous discharge of the motor cells of the antagonist.*

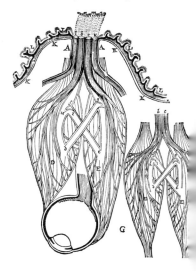

304

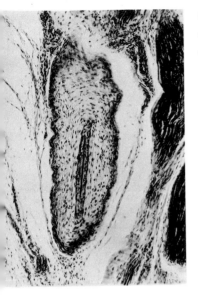

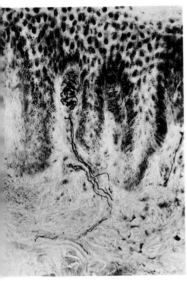

Cutaneous sense organs. The characteristic structure of a given type of sense organ 'tunes' it to a particular waveband of stimulus. Once stimulated, however, it 'broadcasts' a nerve impulse which has no 'qualitative' features at all. The quality of a given sensation is determined by the place in the brain when the impulse arrives.

attachments in the spinal cord. On each side of the spinal cord, therefore, there is a twig through which sensory fibres enter the central nervous system and another one through which the motor fibres leave, an arrangement which is repeated segment by segment, right and left, up and down the length of the spinal axis.

This segregation of nervous traffic is now called the Bell-Magendie Law. Its discovery was the first significant step towards the understanding of nervous action, and in the years that immediately followed scientists succeeded in cataloguing a large number of so-called reflex responses. It soon became apparent, however, that the behaviour of animals was not simply a sequence of reflex twitches, which followed one another whenever the appropriate stimulus was applied. There was a self-evident wholeness about behaviour, an integrated entirety, which was more than the sum of its reflex components. Reflexes collaborated and conflicted, separate stimuli which required the same sort of response evoked a single reply which satisfied both; but if two stimuli were to demand contradictory or conflicting responses, priority was invariably given to the stimulus which represented the more important threat or gratification, and the outcome was invariably functional. Between the entrance and the exit some sort of creative interaction seemed to take place: the creature was not a bundle of reflexes; it was a republic of them. To all intents and purposes, the nervous system behaved in a grammatical fashion, uttering not isolated words but whole sentences and paragraphs of biologically significant behaviour. The formal analysis of this grammar was made by one man, Sir Charles Sherrington.

The most famous clinical reflex is the knee jerk: anyone who has been examined by a doctor must have undergone this test. The familiar kick in response to a blow on the knee first attracted scientific attention in 1875, and there was an immediate controversy about what sort of event it was. Some thought that it was a local response, confined to the muscle itself, but Sherrington insisted that it was yet another example of reflex action, and that it therefore involved a sensori-motor loop through the spinal cord. He used the Bell-Magendie Law to prove this, showing that he could abolish the reflex by severing either the sensory or the motor root. But where did the sensory impulse arise? What sense organ responded to it?

In addition to the five traditional senses, scientists have always vaguely assumed the existence of a sixth, a so-called 'muscular sense' capable of appreciating mechanical tension. The fact that one can move one's limbs with great accuracy in the dark proves that there must be something in the muscles that conveys information about their position. By the end of the nineteenth century, microscopes had improved enough to show the presence of small, spindle-shaped structures embedded in the muscles, and in 1894 Sherrington proved that these were linked to the spinal cord by nerve fibres which entered it through the sensory root. As a result of delicate experiments performed in England and Sweden after the Second World War, we now know that these spindles work like spring-balances, recording the tension of the muscles in which they are embedded. According to Sherrington, the blow on the knee-cap exerts a sudden stretch of the great

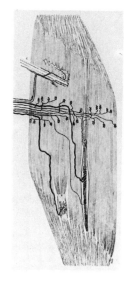

The discovery of the muscle spindle laid the material foundation of a sixth sense. Without these sense organs we would be unable to appreciate the position of our own limbs.

As far as Bell was concerned, the reflex arc was an unbroken 'nervous circle'. This arrangement explained the consistent repeatability of each reflex in isolation but it did not account for the way in which reflexes interacted with one another to create integrated gestures.

With the microscopic discovery of discrete nerve cells it was eventually recognised that the reflex arc was interrupted by at least one anatomical gap. Without knowing how nerve impulses crossed this gap, Sherrington recognised that the interval or synapse was the creative junction of the nervous system — the point at which neurological options were established. Sensory messages might 'escape' to other circuits at this point.

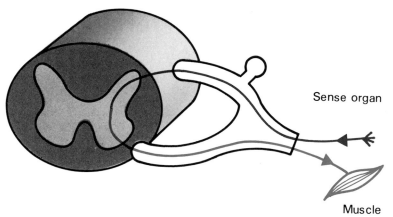

Sense organ

Muscle

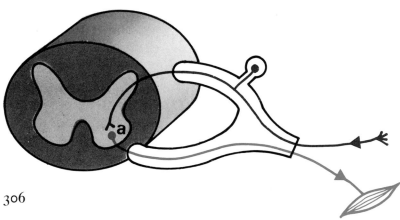

Sir Charles Sherrington established a syntax or grammar of nervous action. He recognised the reflex arc as the simplest intelligible sentence and then went on to show how these sentences were composed into paragraphs and chapters of fruitful action.

extensor muscle of the thigh and, since the spindles are arranged with their long axes parallel to the muscle fibres in which they are embedded, they are lengthened by the same amount and made to fire a volley of sensory impulses into the spinal cord. The volley then ricochets into the appropriate motor channel, and the thigh muscle shortens, taking the strain off the spindles.

This is the simplest observable reflex, and because it enters and leaves the spinal cord at the same level Sherrington called it a 'short spinal reflex', to distinguish it from reflexes where the sensory impulse enters at one level only to give a response elsewhere. Simple as it was, however, Sherrington recognised that there was an important complication. In order to allow the unimpeded contraction of the muscle on the front of the thigh, there had to be an equal and opposite relaxation by the one on the back. As the muscle on the front shortens, its mechanical opponent lengthens by the same amount. Sherrington proved, however, that this co-operative relaxation is not simply a passive surrender to *force majeure*, in which the strength of one contraction overwhelms that of its antagonist. He was able to show that the muscle on the back of the thigh — the one which bends the knee — is actively inhibited at the

The successful performance of a task like this involves the close co-operation of information provided both by the eye and by the working muscles. Vision allows the workman to anticipate the effort which might be needed but the muscular sense monitors the strain from one moment to the next.

very moment the muscle on the front of the thigh goes into action. In other words, although the knee jerk is the simplest conceivable reflex, it involves two simultaneous processes: the active excitement of the muscle which executes the movement, and the equally active inhibition of the muscle which would otherwise oppose it. Sherrington was able to show that this principle applies to all muscular movement. Our muscles are arranged around the joints in mutually antagonistic pairs: a muscle which bends the joint in one direction is invariably opposed by one which moves it in the other. Efficient movement, therefore, involves a balanced process of give–and–take on each side of the joint. Sherrington called this process 'reciprocal inhibition', and insisted that it was the basic grammatical form of all muscular action. It is equivalent to the function of the part of speech which we know as 'but'.

How is this reciprocity achieved? How can a muscle be made to slacken when it is already inactive? The point is that a muscle which yields to reciprocal inhibition is not nearly so inactive as one might suppose. All muscles in the living animal are in a state of mild contraction, even when they are not producing observable movement. In fact, the maintenance of a static posture is just as active and just as strenuous as the movements that arise from it. In order to stand, for instance, the muscles which oppose one another on either side of the knee, hip and ankle respectively must exert an equal and opposite tension. This stabilises the joints and prevents them from folding up under the animal's weight. Strictly speaking, movements are simply the modification of posture, taking the creature from one stable stance to the next: we re-sculpt ourselves from moment to moment. Sherrington called this postural activity 'tone', and the contractions responsible for it 'tonic contractions', a concept which immediately makes sense of reciprocal inhibition. When an animal moves one of its limbs, it increases the state of contraction on one side of the joint and subtracts or extinguishes a proportional amount on the other.

Sherrington realised that to explain how this process of subtraction is accomplished the traditional picture of the reflex arc had to be modified. If each arc was an unbroken circle, looping uninterruptedly from a given muscle, there was no way in which the process of inhibition could occur: the

Human action is a question of melody and harmony. Each muscle has its own melodic sequence but the melody of every individual muscle has to be simultaneously harmonised with the tunes which are being played on all the others. The spinal cord is the organ which integrates these two aspects of physiological function. The brain conducts and orchestrates the larger themes, composing them in the light of far-reaching considerations.

By 'staining' his sections with silver salts the Spanish scientist Ramon y Cajal successfully visualised the anatomy of nerve cells, proving above all that each cell was an independent unit, in touch with others but not continuously linked to any of them.

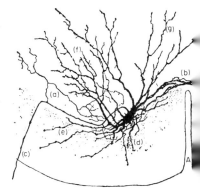

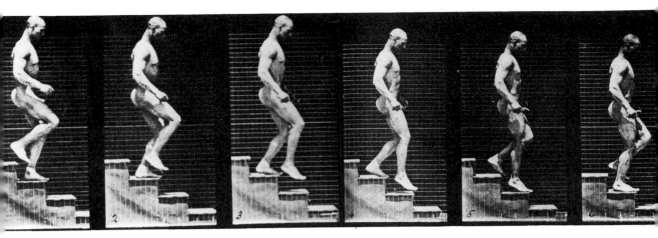

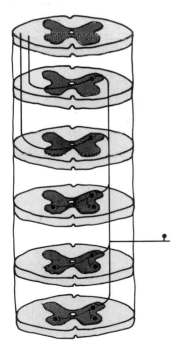

Sherrington's microscopical research showed that the sensory nerve made much more extensive connections than anyone had previously supposed. A stimulus could therefore 'irradiate' or spread well beyond its own spinal segment and could thus call upon the services of relatively distant motor cells.

stimulus would simply race round the circle and cause a contraction without being able to escape and extinguish the muscular tone of its opponent. When he examined the microscopic structure of the spinal cord, however, Sherrington found that the unbroken circle was a fiction. Apart from the fact that it was interrupted by gaps, or 'synapses', the layout was much more complicated than his predecessors had suspected. On entering the spinal cord, each sensory fibre branched in many directions. The incoming stimulus entered a forest of alternative pathways, only one of which led towards the active muscle. As far as Sherrington could see, the volley of sensory impulses was distributed among all these alternative pathways. Even in the case of the simplest reflex, the knee jerk, there was a double distribution of the incoming sensation: one part of the volley looped back to cause the muscle in the front of the thigh to contract, whilst another part exerted an inhibitory influence on the nerves which were already causing tonic contractions in the muscles on the back of the thigh.

Very few reflexes are confined to a single pair of muscles, as they are in the knee jerk. Sherrington found, for example, that when certain areas of skin — the palms of the hand, the soles of the feet, the skin of the crotch, face or head — were superficially stimulated the animal responded not with one pair of muscles but with the widespread and harmonious interaction of several pairs simultaneously. If a dog's ear is tweaked, it

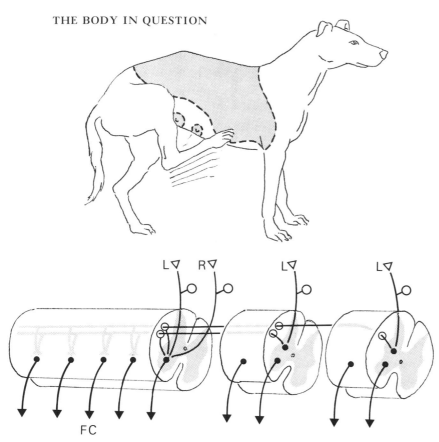

The sense organs capable of eliciting the scratch reflex are widely distributed over the back and flank. The fibres from this region converge on the final common pathways responsible for moving the muscles of the hind leg.

L▽   R▽   L▽   L▽

FC

Each motor neurone receives nerve impulses from many different sources. Some of these tend to excite it into action, others tend to reduce or inhibit its activity. At any given moment it is receiving both excitation and inhibition, and its behaviour depends on which of these has the upper hand. All these impulses are delivered through small nervous termini called boutons terminaux.

turns its head away and at the same time brings up its forepaw, as if to remove the irritation. A stimulus applied anywhere within a large, saddle-shaped area over the dog's back and neck produces an even more complicated response: the hind limb on the affected side reaches forward and vibrates in the familiar scratch reflex, while the hind limb on the other side braces itself against the ground in order to support the animal. Reflexes like this can be obtained in animals that have had their whole brain removed, which means that the isolated spinal cord is able to put together complex and useful harmonies of muscular action.

As it enters the spinal cord, therefore, a sensory stimulus is automatically distributed in two directions: on the one hand, to as many muscles as are needed to compose a useful reply, and, on the other hand, to inhibit those muscles whose continued action would obstruct that reply. 'It has been remarked', Sherrington wrote, 'that Life's aim is an act, not a thought. Today the dictum must be modified to admit that, often, to refrain from an act is no less an act than to commit

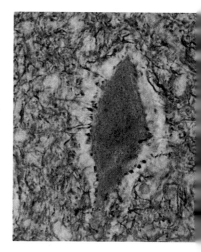

310

one, because inhibition is coequally with excitation a nervous activity.'

Not all the reflexes Sherrington demonstrated were self-evidently useful. He was able to prove, however, that the responses could always be interpreted as the expression of some useful purpose, and if they fell short of this it was because they had been abbreviated or distorted by the experimental situation — by injury, say, anaesthesia or unnatural forms of stimulation.

It was not simply superior intelligence that enabled Sherrington to recognise this: a recent scientific theory had made him sensitive to signs of biological purpose. Nineteenth-century biologists were often reluctant to think in terms of purpose at all, since it re-introduced discredited notions of psychic urge or motive on the part of living matter — something from which the scientific community was now eager to escape. With the appearance of Darwin's theory of evolution, however, the idea that an organ or a function could be described in terms of the purpose it fulfilled no longer raised the

Sherrington identified the motor neurone as the final common path, *the last point at which impulses arising from many different sources could interact and bring about a reflex response. Beyond that point the outcome was irreversible, as the output was inaccessible to further influence. The motor neurone, therefore, is the essential unit of nervous integration, receiving messages from many different sources, some excitatory, others inhibitory. Its behaviour is therefore determined by the arithmetical sum of these influences. For example, if two impulses arrive on its surface they may* together *succeed in detonating the motor neurone, whereas if they had arrived separately they would have failed — so-called* summation.

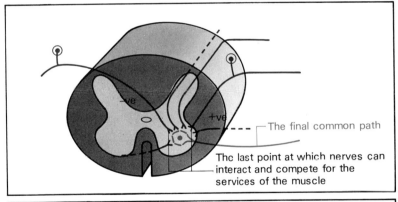

The final common path

The last point at which nerves can interact and compete for the services of the muscle

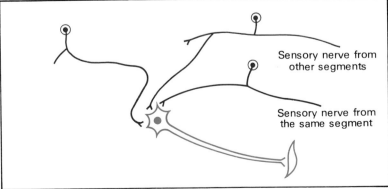

Sensory nerve from other segments

Sensory nerve from the same segment

spectre of mysticism. According to the theory of natural selection, animals acquired useful features not because they strained after them, not because there was a self-conscious urge towards improvement, but because the struggle for existence automatically eliminated those types which were not endowed with adaptive characteristics. In the light of this analysis, it was not only possible but imperative that biologists consider the purpose or usefulness of whatever they found.

Most of Sherrington's work was carried out on animals whose brains he had removed. This simplified the task of reflex analysis by eliminating the higher functions of foresight, memory and intelligence. To all intents and purposes, he created a neurological marionette — a simplified creature which corresponded in many respects to Descartes's 'brute machine'. He was delighted to see what an extensive repertoire remained: even without a brain, the creature seemed able to produce useful responses to the stimuli applied to its skin and muscles; and through the reciprocal processes of excitation and inhibition, the spinal cord exerted what he called its 'integrative function', moulding a total individual from its isolated muscular responses.

Sherrington was not embarrassed by Descartes's theologi-

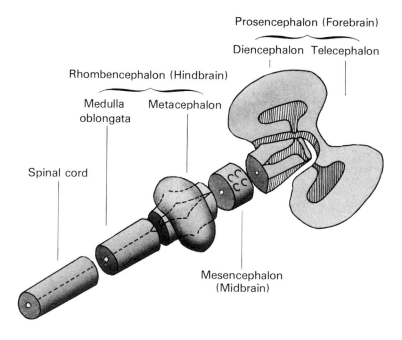

The cerebro-spinal axis. The spinal cord executes the local integrations of trunk and limbs. The higher centres take control of these tactical capabilities and employ them in a strategic context.

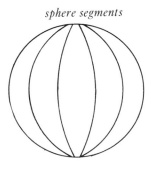

sphere segments

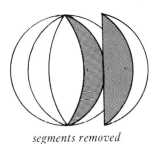

segments removed

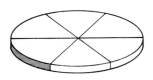

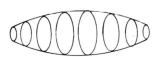

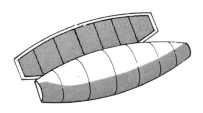

cal scruples. Although the functions of the brain were much more complicated and versatile than those of the spinal cord, he refused to regard this superiority as the expression of a rational soul. What was peculiar about the head was that it was linked to sense organs which brought the creature into relationship with events which had not yet come to pass in its immediate vicinity. The sensations entering the isolated spinal cord conveyed information about the state of the skin and muscles and, since these sensations represented urgent challenges, the response they caused was stereotyped and automatic. The sense organs attached to the brain — the eyes, ears and nose — created a panorama of forthcoming events: they perceived threats and opportunities long before these had a chance to inflict damage or confer satisfaction.

According to Sherrington, then, the nervous system evolved not by acquiring an immaterial soul, but by developing sense organs that introduced the possibility of a prosperous future.

In *The Symposium*, Plato suggested that there was a time when the human body was completely spherical, with the features more or less evenly distributed around its circumference. Plato was nearer to the truth than he knew, for although man himself was never spherical, his primitive ancestors were arranged on a radially symmetrical plan: a cut across any one of their infinite diameters would divide them into identical halves. When an animal is circular, albeit disc-shaped rather than spherical, restrictions are imposed on both its movement and its perceptions.

The broad curves of a circular surface offer mechanical resistance to rapid movement and, since the muscles of such creatures repeat the shape of the rim, their movements are confined to contraction and dilatation: the bell-shaped jellyfish can only pulse and drift; the shallow cylinder of the sea anemone is rooted to the spot and can only retract and expand. When one's repertoire of movement is as limited as this, there is no particular advantage in having an extended field of perception, either. Without the ability to perform nimble movements, it is impossible to take advantage of events which are recognised from afar: there is no point in appreciating a remote threat if you can't flee from it; there is no such thing as a distant opportunity if you can't leap forward and seize it.

The sense organs of such sluggish creatures have a simple structure and a short range; they live in a world where both time and space are equally featureless: the world around is a foggy circumference of unappreciated possibilities. Existence, such as it is, takes place in an infinitely extended now, a totally restricted here. All events are sudden.

One of the most significant landmarks in zoological history, therefore, was the appearance of a new model. At a comparatively early stage in the evolution of life, elongated forms appeared, with features arranged symmetrically on either side of the midline. Even primitive worms have a right and left: they will divide into two equal halves only if they are vertically sliced from end to end.

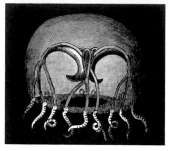

Once an animal is organised in this way, it can slip rapidly through the water along the line of its own axis, and there is an obvious advantage in leading with the same end on every occasion. Such creatures are distinctly orientated in all three dimensions: gravity gives them a top and a bottom; bilateral symmetry gives them a left and a right; leading with the same end on every occasion gives them a head and a tail. They have not only an attitude but an approach to the world, and their instruments of action and perception modify themselves accordingly. The rear end becomes an organ of locomotion, whilst the front becomes the organ of feeding and finding. The mouth appears, and around it are grouped long-range sense organs which pick up the sights, sounds and smells of the world ahead.

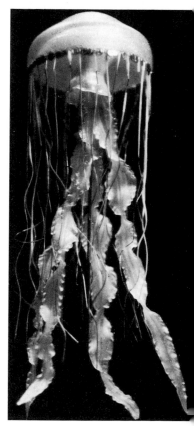

As evolution progresses, the predominance of the front end becomes more and more apparent. It is already identifiable in simple marine worms, whose prows are adorned with elementary eyes and flexible antennae. In crustacea and insects the development is even more pronounced: the limbs in the forward segments have now lost their locomotor function and are grouped around the mouth to form specialised jaws and feelers; there are multi-faceted eyes equipped with lenses, and these can give clearly focused images of distant objects. By the time land vertebrates appear, the sense organs have become enormously sophisticated, and the head is mounted on a mobile neck, so that the whole module has become an elaborate instrument of exploration and anticipation.

Along with these developments, the front end of the central

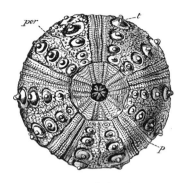

When an animal chooses to approach the world with one end rather than the other, the segments at that end acquire special instruments of action and perception, and the rear end becomes an organ of propulsion.

nervous system becomes distinguishable as a brain. In the trunk and tail the nervous system consists of a chain of nodes or ganglia which co-ordinate replies to local events on the body surface. The nodes of the leading segments, however, clump together to form a clearing-house for the specialised information coming through the long-range sense organs and, since this information puts the creature in touch with its situation as a whole, the head inevitably assumes mastery of the rear end: the powerful muscles of the limbs and trunks are put at the service of the intelligence agency situated up front.

The brain is the great initiator: it proposes and inaugurates. But its initiatives have to be realised in terms of co-ordinated muscular action. The spinal cord acts as the distributor, and all voluntary action is therefore abolished when the connection is severed. However, as Sherrington's experiments showed, the spinal cord is not just a passive cable: it has a repertoire of its own, and decapitated primitive vertebrates are able to perform quite complicated acts of locomotion. For instance, a frog with its brain removed can swim — and until one examines its behaviour more closely and discovers that it is swimming to no good purpose, it is easy to get the impression that the injury is trivial. The repertoire of a decerebrate dog is much more seriously curtailed: it has a rigid, statuesque posture, and, although it can be be propped upright on all fours, it can't adjust its stance, and topples over at the slightest touch; it can manage isolated fragments of standing and stepping; it will flinch and withdraw its foot from a painful stimulus, and it can give quite a creditable performance of scratching, but all these actions are piecemeal and crude. They are the actions of a derelict robot. In man, the situation is even worse: so much control has been handed over to the brain that when the connection with the spinal cord is severed by illness or injury the repertoire is reduced to a few crudely generalised reflexes.

However, the paralysis associated with brain injury is quite different from that which follows local injury to the spinal cord itself. The powerlessness of a patient suffering from a stroke is not the same as that of a patient paralysed by polio. If you manipulate the limbs of someone who has been paralysed down one side by a cerebral haemorrhage, you will notice a pronounced rigidity. The patient may be unable to move his

315

joints at will, but he puts up an automatic resistance to having them manipulated for him. As you bend and straighten the joints, you can almost feel the muscles tensing against your efforts, and, if you strike the various tendons with a reflex hammer, the responses are much brisker than they are on the normal side. The paralysis is spastic; in other words, in a state of spasm. When these patients begin to walk again, the affected foot is not simply dragged along the ground, but seems to be actively pressed against it. If you look at the shoe of such a patient you often find that the toe has been rubbed away as a result of the unusual friction.

All this is in sharp contrast to the paralysis. which results from injury to a peripheral nerve or from local infection of the spinal cord by the virus of poliomyelitis. Then, the muscles are flabby and show no sign of tone. The reflexes are diminished, if not entirely absent: it is impossible to obtain the familiar tendon jerks. The paralysis is characteristically flaccid.

In other words, when the inaugurating apparatus of the

The brain is the pilot or inaugurator of biological action.

John Hughlings Jackson. Under the influence of the evolutionary thought of the English philosopher Herbert Spencer, Jackson visualised the nervous system as a hierarchy of organisational levels, each one suppressing and modulating the activities of its primitive predecessor. Injury or disease destroyed or toppled this edifice and brought about a crude recapitulation of evolutionary history. (Institute of Neurology, London)

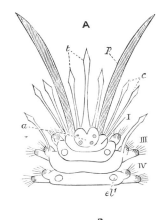

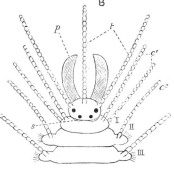

brain is damaged, the paralysis has an active component, and this is missing when the distributor is knocked out. This implies that the brain exerts a restraining influence on the native spontaneity of the spinal cord and that, in addition to initiating actions of its own, it somehow withholds and modifies those of its obedient servant.

One of the first scientists who tried to account for this was the nineteenth-century English clinical neurologist John Hughlings Jackson, who spent most of his life observing and cataloguing the behaviour of patients suffering from strokes and epilepsy. As he watched these patients, Jackson saw that their symptoms invariably sorted themselves out into two groups — positive and negative. On the one hand, the patients lost certain abilities. They were paralysed down one side and unable to perform voluntary movements: following epileptic seizures, certain patients passed through a brief phase of one-sided paralysis similar to that following a stroke. Sometimes they lost the power of speech. On the other hand, Jackson noticed that as long as the patient survived certain nervous functions were retained, and these residual functions —crude muscular reflexes like the knee jerk or the automatic emptying of the bladder, for example — were much simpler and much more automatic than the ones that had been lost.

Jackson insisted that these patients had reverted to an ancestral condition, that the illness had stripped away higher levels of nervous-system development — that is to say, those most recently acquired — leaving the older ones exposed — rather like an archaeologist exposing more and more ancient remnants as he deepens his dig. He wrote:

I have long thought that we should be very much helped in our investigation of diseases of the nervous system by considering them as reverses of evolution. By evolution I mean a passage from the most simple to the most complex, a passage from the most automatic to the most voluntary. The highest centres, which are the climax of nervous evolution, are the most complex and the most voluntary. So much for the positive process by which the nervous system is put together. Now for the negative process, the taking to pieces or dissolution. Dissolution is a process of undevelopment; it is a taking to pieces in order from the most complex and most voluntary towards the most simple and most automatic. I have used the word 'towards' for if dissolution were total the result would be death. But wherever dissolution is partial, evolution not being entirely reversed, some level is left. There is thus a negative and positive element in every case.

But the disease had not simply revealed lower functions: it had somehow released them as well, allowing them to express themselves with unprecedented vigour. These are the spastic symptoms that I have just referred to. It was as if the brain had two functions: the ability to bring about voluntary movements, and the equally important ability to moderate involuntary reflexes.

The higher nervous system not only enlarged the repertoire of the living organism, Jackson noticed, but also appeared to exert a civilising influence on the more primitive part. He found, for example, that patients who had lost most of the power of speech were sometimes left with an exaggerated tendency to utter oaths and swear-words; patients with brain damage sometimes became energetically uncouth.

This led him to believe that evolution was not simply the successive addition of more and more sophisticated skills, but the concomitant repression of the more ancient ones: the evolution of the nervous system did not proceed like the building of a pagoda, piling more and more elaborate tiers one on top of another; it was a succession of restraints as well, pressing down the lid on the jack-in-the-box of all previous evolutionary stages. Illness or injury brought about a recapitulation of the patient's evolutionary history: by losing abilities that his species had recently acquired, he was forced to act out in an exaggerated way the repertoire of his forgotten predecessors.

Jackson's doctrine of regression was so persuasive that many of his colleagues tried to extend the notion too far. In 1903 the English anthropologist W. H. Rivers and the clinical neurologist Henry Head tried to apply Jackson's theory by examining the effects of an injury deliberately inflicted on one of their own sensory nerves. On 25 April 1903 Henry Head submitted himself to a surgical operation performed by one of

When the highest level of motor function is destroyed — by haemorrhage, thrombosis or injury — the patient suffers a 'stroke' on the opposite side of the body. The muscles and their spinal nerve supply are intact, but the patient has no access to them. She cannot initiate voluntary movement. She has lost the great inaugurator on that side. Such patients do not simply fail to move their limbs — they fail to try and move them. They cannot conceive what would count as trying.

his colleagues at the London Hospital. A small sensory nerve in his forearm was carefully cut; the ends were re-joined and stitched with fine silk. The wound was closed, the limb splinted, and the hands left free for testing. The incision healed without further incident. On the following morning, a large patch on the back of the hand and thumb was found to be quite insensitive to pin-prick, light brushing with cotton wool, and all degrees of heat and cold; there was no sensation when the hair on the skin was tweaked. The severed nerve was no longer carrying sensory messages to the central nervous system; but, since the nerve had been carefully stitched together, its fibres were able to regrow and establish their old connections, and during the long period of recovery Head and Rivers were able to chart the return of sensation.

One might have expected recovering sensation to encroach from the edge — rather like a puddle drying up in the sun. And as far as the shape of the patch was concerned, this was the way it went. Head insisted, however, that the recovery of sensation came in two phases. To begin with, feeling began to encroach on the anaesthetic area. But the *sort* of feeling was very unusual. It had a much higher threshold than the skin on his other arm — Rivers had to push much harder to produce any sensation at all — and once the threshold was crossed, the sensation was unpleasantly intense and hard to locate accurately: it had a coarse, primitive roughness which they called 'protopathic'. When the whole patch had recovered some sort of sensation, Head began to notice a second phase of recovery — as if the crude sketch was now being shaded in and coloured. The threshold of sensitivity dipped to normal levels — it was possible to obtain sensations with comparatively light stimulation — and the feeling was discrete, localised and informative. This final stage of recovery produced what they called 'epicritic' sensation.

According to Head and Rivers, this little experiment brought about a brief replay of nervous evolution. In the first phase of recovery, the primitive, coarse, ancient nervous system was revealed in its true colours — released from the inhibition of the more sophisticated stages, it expressed itself with unrestrained vulgarity; as the nerve was restored to full function, the dog beneath the skin was restrained and put back on its leash.

These experiments have been repeated several times, and no one since has been able to get the same results. In any case, the interpretation seems very shaky. Quite apart from the fact that the observations were unreliable, since they were reported by someone whose own interests were involved, the conclusions do not really make sense. It seemes highly unlikely that the ancestral nervous system was as crude as Head and Rivers maintained — only a sea anemone could hope to prosper with such sensitivity. The point is, I think, that they were unconsciously interpreting their own findings in a metaphorical rather than a scientific manner: they had started out strongly biased in favour of Jackson's evolutionary theory of nervous illness, and their convictions were so firmly held that Head at least unconsciously reshaped his own feelings until they confirmed the theory which had moved him to embark on the experiment in the first place.

This is not to say that theoretical preconceptions, as such, are any bar to success in science. It is impossible to imagine what else would give a scientist either the interest or the incentive to conduct a scientific experiment. The preposterousness of the stereotype of the scientist as a disinterested observer — someone who approaches the world with an open mind and a sensitive eye — whose duty is to amass as many facts or truths as he can, filing them away as raw material for the manufacture of something called a theory, is illustrated by the story of the man who spent his life collecting facts and observations, which he bequeathed to the Royal Society in the hope that the distinguished Fellows would commemorate him by shaping his collection into an interesting theory.

Observations do not litter the world like seashells or butterflies, waiting to be picked up by an empty-headed naturalist. They are made by someone who has a vested interest in the particular range of phenomena, and what gives him this interest is a hunch about the way in which the world works. This attitude has been expressed most eloquently by Karl Popper: 'Observation is always selective. It needs a chosen object, a definite task, an interest, a point of view, a problem.' Popper points out that the observations a scientist makes are determined by the theories he accepts as a background. In other words, all science starts with an act of conjecture, and

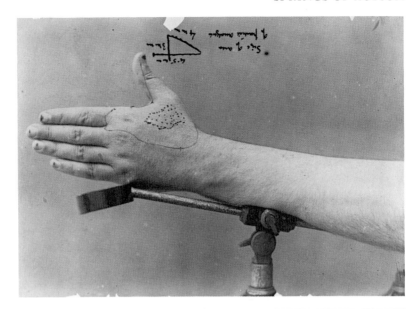

Head and Rivers conducting their classic experiment in Rivers's room in St John's College, Cambridge.

observations and experiments are designed to test this to its breaking-point: the truth of a scientific theory is proportional to the amount of falsification it has endured and survived.

What makes Head's and Rivers's conclusions so ambiguous is not that they undertook their experiment in the light of a persuasive theory, but that they failed to acknowledge the contradictory evidence. They were not dishonest: it is hard to imagine a more scrupulous report — the daily observations fill several carefully annotated volumes. But the very form of the experiment created pitfalls. Only one person had access to the relevant observations and, since cutaneous sensations are notoriously susceptible to the subject's own state of mind, it is almost impossible to imagine how the experiment could have turned out as anything other than a corroboration of the theory which prompted it.

Why was Jackson's theory so persuasive, though? It is partly because it seemed to follow so neatly from Darwin's. Like Newton in the previous century, Darwin had given the scientific world a doctrine of unprecedented scope and power, and scientists whose work was not yet covered by the theory hurried to reorganise their research so that they, too, could sail under these impressive colours.

But, I wonder whether this is the whole explanation. Although Head and Rivers were quite honest in the lip-service they paid to Jackson's Darwinism, I am convinced that there were other less openly acknowledged forces at work — forces which probably influenced Jackson himself. When Jackson summarised his doctrine of release, he expressed himself in revealing political terms: 'If the governing body of this country were destroyed suddenly, we should have two causes of lamentation: 1. the loss of services of eminent men; 2. the anarchy of the now uncontrolled people.'

The idea that civilised man has a savage inside him is, of course, as old as civilisation itself and, long before Darwin, Thomas Hobbes had insisted that the aggressive appetites of individuals could be reconciled only if everyone submitted to the restraints of a single sovereign authority. Left to itself, nature was in a state of war, and man unregulated by society would eke out an existence which was 'nasty, brutish and short'. Jackson lived at a time when Hobbes's theory had been dismally confirmed by the events of the previous 100 years.

Plato compared the human passions to a spirited horse reined and controlled by the sovereign power of Reason. In the era of Romanticism and revolution this image was revived and celebrated. (David, 'Napoleon crossing the Alps', Versailles)

The spectacle of the revolutionary mobs of 1789, 1830 and 1848 reminded European intellectuals that human beings were restrained and civilised only by their institutions, and that man could easily return to a state of barbarism if these artificial restraints were relaxed. The idea of regression must have been just as captivating as the biological theory of evolution. By demonstrating that man was genetically related to the lower orders and that he retained active residues of his own primitive ancestry, Darwin made it easy to believe that the tendency to regress was not an accidental misfortune, but was written into the very constitution of man.

Writing at exactly the same time as Jackson, the political economist Walter Bagehot coined the resonant term 'atavism':

> Lastly we now understand why order and civilisation are so unstable even within progressive communities. We see frequently in states what physiologists call atavism. The return in fact to the unstable nature of their barbarous

ancestors. Such scenes of cruelty and horror as happened in the French Revolution and as happened more or less in any great riot, have always been said to bring out a secret and repressed side of human nature. And we now see that they were the outbreak of inherited passions long repressed by fixed custom but starting into life as soon as that repression was catastrophically removed.

The increasing success of the doctrine of regression probably owed as much to straightforward social pessimism as it did to the theory of evolution, and if Head overinterpreted his own findings as examples of regression it may have been as much the result of political anxiety as it was of enthusiasm for evolutionary theory.

But this, too, would not on its own discredit his results. His conclusions would have been just as shaky if the sources had been purely Darwinian. The point is that the results were unrepeatable, biologically implausible, and suggested an entirely unrealistic relationship between one part of the nervous system and another. The damaged nervous system is bound to be less efficient than the intact one: both action and sensation will inevitably deteriorate when their physical foundation is injured. But there is no justification for saying that the repertoire of the damaged nervous system is a replica of any one of the previous healthy states.

This creature may have looked before it leapt but the up-dated information provided by the balancing organ in the ear and the muscle spindles in the legs keep the performance on line from start to finish.

By the same token, the fact that the brain is a more recent acquisition than the spinal cord does not mean that when it is damaged the functions that are set in action express those of the spinal cord as it was in the days of yore. The nervous system does not evolve by a successive addition of parts which leaves all the rest in a state of arrested development. The acquisition of something new is accompanied by a progressive modification of everything that was there in the first place. The spinal cord did not stop evolving because more sophisticated apparatus was being screwed on to the front end: it was becoming more sophisticated itself at the same time, so that damage to the more recently acquired parts of the nervous system would not automatically re-create a picture of some previous stage of evolution. In any given creature, man or dog, the brain and spinal cord are contemporaries and, although their co-operation may involve inhibition of the latter

by the former, to suggest that this demonstrates the power of the new over the old is like saying that the brake pedal on a car exercises a sophisticated restraint on the primeval energy of the engine.

Admittedly, most of the interesting initiatives that characterise an individual originate in his brain and are merely executed and distributed by the spinal cord. But the traffic is not all one way: the spinal cord distributes just as much information to the brain as the brain broadcasts in the other direction. In fact, the more evolved the nervous system is, the richer this reciprocity becomes.

The strategic estimates which the brain compiles on the basis of information coming from the eyes and ears can result in profitable actions only if such information is discussed and checked against equally relevant information about the tactical resources of the limbs and trunk. Without such cross-reference, ambitious schemes conceived by a far-seeing brain would remain unfulfilled fantasies. Feeling thirst, for instance, you might turn your head and swivel your eyes until your glance comes to rest on a glass of water. The brain may improvise a clear strategic plan of what to do with the glass of water, but the tactical problem of lifting it from the table to your lips can be solved only by moment-to-moment estimations of how the lift is going. If the grip is too strong and is not automatically limited by the experience of pressure felt by the fingertips and the joints of the hand, the glass may splinter. If the grip is too weak, because the slippery surface has not been anticipated, the glass will fall to the floor. And that is only the start of the problem. The eyes may give some idea of an object's weight, but vision is notoriously deceptive. The fact that the glass is raised to the lips without being smashed into the face is a tribute to the subtle weighing abilities of the outstretched limb. And the fact that the glass remains at the mouth while losing weight as it is emptied shows how punctually the news is updated: without this information the glass would levitate as it was drained.

The sensations coming from the joints and muscles are so unconsciously heeded that it is easy to overlook their importance in comparison with the vivid experiences of sight and sound. However, when illness interrupts the flow of information from the muscles to the brain the patient finds it very

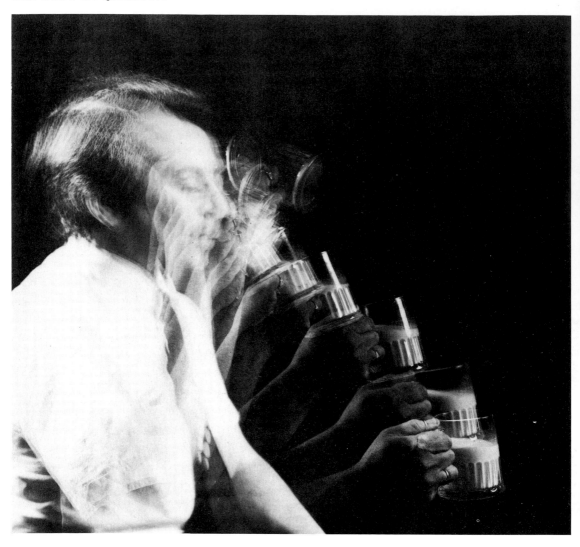

hard to make sensible tactical decisions. This is what happens to people suffering from neuro-syphilis. In this illness, which is now quite uncommon, the patient loses muscular sensations from his lower limbs and is forced to over-rely on visual information. The result is that when he closes his eyes he falls over. Such patients are so cut off from information about their lower limbs that they find it hard to co-ordinate their gait. Since they are always in danger of falling over, they try to make up for the shortage of updated news by walking with their legs as wide apart as possible. And since normal stepping no longer brings back adequate sensation from the joints and tendons of the feet, the patient slaps his feet down hard at each step in an effort to bang the information past the block.

Muscular performance is constantly being reshaped by information about its success or failure. As the action progresses the sensations which result enter the brain, which in turn adjusts its output in the light of what it has just been told. If this feedback is distorted *en route*, the performance goes haywire. It may result in a muscular overshoot. If this overshoot is misreported, the movement which is ordered as an adjustment may itself overshoot, and so on and so on, back and forth, getting worse and worse. This results in a tremor, which gets more exaggerated as the limb approaches its target.

The stability of muscular performance is yet another example of the all-pervading influence of negative feedback. The relevance of this principle was vaguely appreciated but not explicitly stated as soon as the existence of muscular sense organs was established at the end of the nineteenth century. It was only in the Second World War, however, that the system was freely compared with mechanical devices which exemplified the engineering device known as a servo-loop. Once again, this was the result of biological scientists unexpectedly coming into contact with a rapidly developing technology.

In the early years of the war, doctors and physiologists, both in England and America, were seconded to military establishments in the attempt to improve the performance of pilots, gunners and bomb-aimers. In order to evolve efficient manual controls, it was necessary to analyse the neuro-muscular characteristics of the human operator, treating him as a mechanical component of a self-regulating system. In each of the tasks that were studied, the purpose of the exercise was to maintain a steady state, with as few errors or fluctuations as possible. In the aiming of weapons, for instance, the gunner was supposed to follow the moving target, keeping it on or close to the cross-hairs of his range-finder. What sort of controls would help him to achieve this? How much resistance or friction should the wheels or levers have? What was the most favourable gearing ratio? By investigating these factors, by varying them one by one, and observing the improvements in aim and performance, the efficiency of mechanical warfare improved out of all recognition.

Shortly after the war, when the medical men returned to

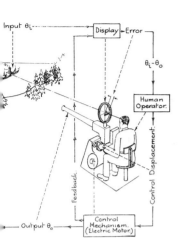

Artillery is simply human action carried on by other methods, and by studying the mechanical extensions of the living body scientists gained an unprecedented insight into the behaviour of the human nervous system.

their civilian work, they realised that they had created a revolutionary model for thinking about muscular performance in general, and that the tracking tasks which they had analysed in a military context were examples of what the nervous system was called upon to do throughout its life. The pursuit of an aerial target is error-activated. The perceived mis-match sets in motion the actions which cancel it out. By elucidating the mathematical principles of such error-activated performance, the military physiologists of the Second World War found that they were printing out the engineering specifications of the human nervous system. By treating a man as if he were part of a mechanical set-up, they began to discover the extent to which he as a whole actually was one. It is no exaggeration to say that the collective experience of the human operator in control systems laid the foundations not only for all neurophysiology in the late twentieth century, but actually created the science of cybernetics.

The sensations which arise from muscle, joint and tendon are conveniently grouped under one heading not because of their anatomical source, but because they collaborate to provide the brain with a distinctive form of information. Sherrington called this 'proprioceptive sensation'. It tells the creature what it is doing and what is happening to it as a result of what it is doing: whether movements are going according to plan, or whether they are being obstructed. In other words, it monitors and re-fashions the creature's muscular enterprises from one moment to the next.

Without such information, there would be no knowing how one's limbs were arranged, and it would be impossible to find one's nose in the dark. The proprioceptive system supplies the brain with a co-ordinated map of all the available muscular resources and their current state of readiness. One of Henry Head's most significant contributions to neurology was his recognition of the fact that the proprioceptive system is largely responsible for providing us with the image of our own active body: he called this image the 'body *schema*'.

None of this information could be put to use unless it was firmly located against a fixed horizon; it would float weightlessly in a vacuum. If you regard the body as an aircraft flying in a thick cloud, the proprioceptive system re-creates a scale-model of the machine on the instrument board: it tells the

After the Second World War medical scientists who had worked for the military recognised that a fruitful analogy could be drawn between the guidance of missiles and the controlled movement of limbs.

Some movements are so rapid that there is no time to get feedback information from the muscles while they are being executed. A rapid musical arpeggio has to be planned in advance, usually as a result of painful practice, and the whole sequence is then launched like a ballistic rocket. During the Second World War experimental psychologists distinguished between movements which were continuously controlled throughout their action and ones which had a ballistic character and could not be modified once they had started.

pilot where the wings are in relation to the fuselage, and whether the engines are developing the right horsepower. It tells him nothing, however, about the aeroplane's overall attitude in space: it could be nose up and tail down, for all he knew, heavily rolled to starboard about its long axis and strongly yawed over to the right. Unless he is kept informed about this, he can't hope to make a realistic approach to the runway. Fortunately, the proprioceptive system is framed and defined by information about the body's overall attitude in space, which is provided by paired sense organs associated with the inner ear. Each of these sense organs acts as a plumb-line. If the labyrinth is damaged or inflamed, the brain may receive a torrent of misleading information: the plumb-line is no longer true, the artificial horizon now bears no relation to the actual one, the patient reels drunkenly and experiences the intolerable sensation of vertigo. It is the subtle collaboration between the labyrinth and proprioceptive system that allows a cat to fall on its feet irrespective of the position it is in when dropped.

The proprioceptive system, labyrinth included, provides the physical basis for the experience and control of action. Sherrington contrasted it with the 'exteroceptive' system, which provided the creature with knowledge about events which were less directly dependent on its own actions: sense organs in the skin create a tight body-stocking of local sensation, while the eyes and ears inflate a balloon of more distant awareness about the creature's head.

But the distinction between proprioceptive and exteroceptive sense organs is less sharp than it seems. It is certainly important to distinguish between sensations which result from one's own actions and sensations which result from the world's actions — sensations which are self-induced and ones which arise uninvited from the environment — but no sense organ ever exclusively reports either one or the other. A receptor which is passively acted upon at one moment may be actively poked in the direction of stimulation at the next. The sense organs on the surface of the body are obviously favourably placed to suffer the accidents of environmental change, but since they are embroidered into the glove of an active limb they can be stimulated just as easily by what the creature does with that limb. Conversely, the spindles and stretch receptors in the tendons and muscles are most frequently stimulated by what the creature does with its muscles — as the biceps contracts and bends the elbow, it automatically stretches and stimulates the spindles embedded in the muscles which would otherwise straighten the elbow; but they can also be stimulated when a passing blow deflects a limb unexpectedly. Proprioception and exteroception, therefore, are not fixed categories of sensation, but alternative modes of getting sensations.

Each type of sensation — visual, tactile or acoustic — is distributed in an orderly fashion to its own particular province in the brain, laying itself out in a pattern which corresponds to the sensory surface from which it is derived, without necessarily reproducing its spatial proportions. We have already seen that the tactile map in the brain is recognisably similar to the body surface which it represents, although its scale is quite different. This is mainly because each part is represented in

The plumb lines of the inner ear register the position of the head in relation to gravity. Inflammation of these organs — labyrinthitis — causes them to broadcast misleading information and the patient reels in the attempt to adjust to the spurious input.

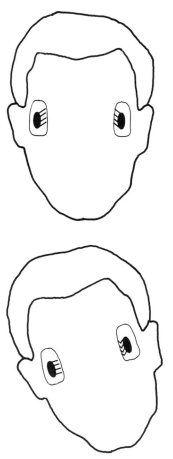

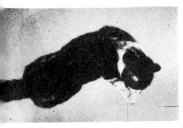

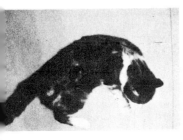

The co-operation between labyrinth and muscle spindles guarantees consistent accuracy of the righting reflexes.

proportion to its importance rather than its anatomical size. The transformation is further complicated by the fact that the picture is projected on to a surface which has a completely different shape. In this respect, it is rather like the effect you get when a magic lantern is thrown on to the heavy folds of a curtain, or like a mercator projection, where the geographical proportions of a spherical world have to be remodelled so that they can be laid out on the flat surface of a page.

But although the image is transformed by projection, the parts are always in their proper place: the shoulder is invariably represented between the trunk and elbow, just as it is in the body; the projected mouth may be larger than the projected foot, but it does not sprout out of the ear. The same geometrical decorum applies to all the other projections. When the retina is re-mapped on the cerebral cortex at the back of the brain, its relative dimensions are re-proportioned, but the spatial relationships are preserved, so that what is north-north-east of a spot in the retina, is still north-north-east when it is recharted in the brain. Within the ear, the acoustic sense organs are arranged in the form of a little spiral, or cochlea — low notes at the bottom, high notes at the top. When this pattern is projected on to the brain, the spiral unwinds and reappears in the form of a straight keyboard, but the notes of the scales follow one another in the same order in which they occur in the ear.

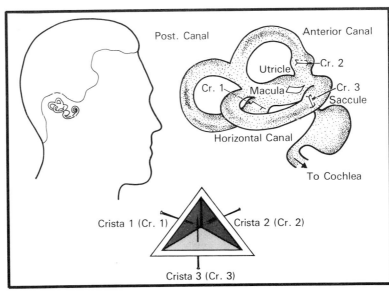

These orderly arrangements were not known about until the latter half of the nineteenth century. The phrenologists, who included among their enthusiastic followers such distinguished persons as George Eliot and Charles Dickens, thought that the brain was parcelled into separate organs, each of which was responsible for one or other of the known capabilities of man. There was an organ of amativeness, one for acquisitiveness, a philoprogenitive organ, and even one for piety. Each of the faculties listed in Roget's *Thesaurus* was thought to have its counterpart in the brain, as if nature had already anticipated the moral distinctions her principal creature would make.

When anatomists and clinicians began to examine the brain more closely, they inevitably discovered that these phrenological provinces were fictions and the various parts of the brain were distinguished from one another according to altogether different principles. One of the first scientists to make this clear was Jackson, during his study of epileptics. He pointed out that the fits were due to the irregular discharge of nerve cells at the site of a tumour or injury. He noticed that in a large number of cases fits invariably started as a small local disturbance — the twitching of a thumb, a tingling at the corner of the mouth, a cluster of scintillating lights in one quarter of the visual field. As the fit progressed, or 'marched', the disturbance spread, so that the patient was first convulsed

The pilot can orientate himself against a visual horizon. In fog he must rely on the artificial horizon provided by his inner ear. The semi-circular canals provide extra information about the rate of acceleration in each of the three planes of space.

throughout the whole of one side of his body, and then on the other, eventually losing consciousness.

According to Jackson, the fit started at the site of the lesion, which acted as a detonator to a spreading cerebral explosion. By correlating the site of the earliest symptom with the seat of the injury as it was eventually discovered at the post-mortem, Jackson concluded that the brain was mapped in an orderly fashion — not in terms of phrenological faculties, but in sensori-motor regions. Injuries, tumours or vascular abnormalities which were found at the rear-end of the brain invariably produced fits starting with optical scintillations; tumours in the parietal lobe were associated with peculiar bodily sensations; and if the patient reported whines, booms or buzzes at the onset of his fit, Jackson knew that he would find a pathological abnormality somewhere in the temporal lobe. By carefully studying these early or premonitory symptoms — the so-called 'auras', the breeze which ruffles the water before the wind fills the sail — Jackson was able to show that the various modes of sensation were independently localised in different parts of the brain, and when surgeons later succeeded in exposing the brain of conscious (locally anaesthetised) patients, stimulating it point by point with electric needles, they were able to demonstrate that many of the provinces of the brain were topographical maps or projections of the sensory fields which they represented.

Unfortunately, the word 'projection' has misleading connotations of pictorial display, which tempt one to assume that the human mind hovers over the dimly lit screen of its own brain like a phantom spectator watching the projected images of the passing scene. But even if one were to accept the notion of a ghostly supervisor over-hearing or over-viewing the input of his own sense organs, it is important to realise that what he would be watching or listening to would bear an indirect and rather complicated relationship to the corresponding events in the outside world. Once a pattern of light has struck the retina, as soon as a train of sound-waves has travelled up the spiral of the cochlea, the events are translated into sequences of nervous information so that the transmission no more resembles what it represents than the currents in a television circuit resemble the pictures which they carry.

The sense organs are not simply windows, listening tubes

Phrenological principles created a pseudo-science of cerebral localisation.

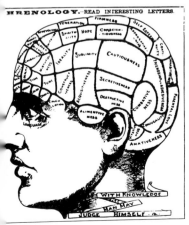

or bell-wires through which the outside world conveys itself in detailed replica to an attentive Lady of Shalott: they are the input channels of a computer which can operate only when the physical events impinging on the sensory surface are transformed into the characteristic digital language of the brain. As these transmissions are handed on from one relay-station to the next, *en route* for the primary projection area, they undergo systematic filtering and readjustment so that even if they could be translated back into visual or acoustic images, the pictures or sounds which would result would be very different from the images which were responsible for them in the first place.

To some extent these adjustments are made necessary by the physical imperfections of the sense organs themselves. Theologians who use the argument from design as a proof of the existence of God frequently refer to the optical perfection of the eyeball. Ironically, this is the one part of the system where the argument from design falters, for the image projected on to the retina is so blurred and unsteady that if one tried to develop an ordinary film from it one would find it almost impossible to reconcile the smeared, hazy print with the crisp, sparkling detail of what one perceives through the eyes. In the early stages of transmission from retina to brain some of these errors are computed out: blurred contours are sharpened and a start is made on preparing the information for the part it will eventually play in conscious perception. In some ways this process resembles the image intensification which enables astro-physicists to build up pin-sharp images of planetary terrain.

The situation is further complicated by the fact that all this information is partitioned amongst the various levels of nerve cell which make up the full thickness of each projection area. At each level there are nerve cells tuned to answer one particular feature of the input and to remain indifferent to all others. Take the visual area. In one layer there are cells which respond to the arrival of light on the corresponding part of the retina; at another level there are cells whose activity is extinguished by the onset of light, but which burst into activity when a shadow falls on the corresponding part of the retina. There are sheets of cells which fire only when a *horizontal* bar of light is projected into the eye, and others which can detect

The anatomical perfection of the eye gives a misleading impression of its optical efficiency. The lens of a modern camera is infinitely more refined. However, the photographic emulsion does not even begin to compare with the versatile sensitivity of the retina. The optical image has been extensively edited and remodelled by the time it is transmitted to the computing machinery of the brain.

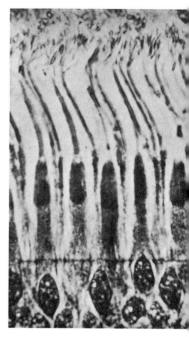

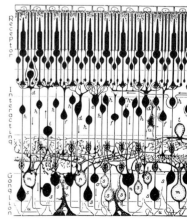

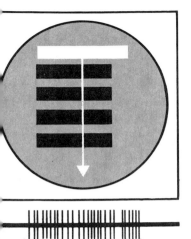

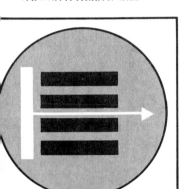

Nerve elements in the visual cortex are not only sensitive to light but to the direction in which the light is moving. One group will fire only when the white bar of light is moving vertically downwards and will remain inactive if the bar is moved horizontally.

only *vertical* stripes. Some cells respond only to certain axes of movement — side to side or up and down.

The projection area, then, is not simply a camera obscura, carrying a miniature replica of what it represents: it is a mathematical analyser which breaks down the physical components of the sensory input into the biologically relevant features of the physical world. Comparable analyses occur in each of the primary projection areas, visual, auditory and tactile.

Such analyses can work only on the information which the sense organs themselves provide — spatial patterns of illumination in the case of the eye and temporal sequences of vibration in the case of the ear. And yet the perceptual experiences of vision, hearing and touch are incomparably richer than that, which suggests that perception is something over and above the mere inspection of all the information which is displayed and analysed in the primary projection areas. What one sees goes well beyond what the eye provides; what one hears is more elaborate and more significant than the meagre information provided by the ear. The visual experience, for instance, is not simply a shifting panorama of contours, shadows and colours. What one sees is a series of enduring objects which are apparently distributed in a coherent three-dimensional space.

Paradoxically, by using one's eyes one is made aware of the non-visual characteristics of what one sees. Hardness, roughness, smoothness, distance and weight, for instance, are not optical characteristics, although one reads them off the optical clues which the retinal image provides. A retinal patch may be large or small, whereas its distance or nearness is something which can be inferred only by 'looking up' one's past experiences with such images and by remembering how they behaved when one walked or moved one's head hither and thither. Impenetrable hardness can be read into a particular sort of appearance only when the optical features are correlated with the memory of tactile resistance. In other words, the sense organs deliver a meaningful world only when they are allowed to transfer some of its resources into the current account of another sensory modality.

The difficulty is that in normal perception these interactions and 'lookings up' are so automatic that the owner of

335

It is tempting to regard perception as a spectacle watched by a phantom observer, as if the brain were a camera obscura *housing the mind's eye. An eye within an eye.* (International Museum of Photography, Rochester, New York)

a brain is quite unaware of the disparity that exists between the information which his brain provides and the images which his conscious mind experiences. However, one can gain valuable insight into the extent of this conjectural activity by deliberately simplifying the sensory input and seeing how much the perception goes beyond what is given.

One of the most vivid examples of this is the creation of so-called 'subjective contours'. In 1955 the psychologist Kanizsa prepared a figure in which a triangle broken in three places was surrounded by three circular black patches, each one of which was interrupted by a white wedge. When one inspects this figure it is almost impossible to escape the impression that the circular patches and the triangle are overlaid by a whiter-than-white triangle. It is evident from this illusion that the brain has gone well beyond the optical information which is provided by the eye and that the subjective contour of the

white triangle is imposed upon the figure as the most satisfactory explanation of the otherwise puzzling discontinuities in the other components. In other words, what the mind sees is not what there is, but what it supposes there might be. And in order to create this supposition it must refer to information which is not immediately given in the figure itself. It can arrive at such a conclusion only by remembering some of its other experiences in which the manipulation of real objects showed that things can stand in front of one another and that when certain figures which have a well-established reputation for completeness seem to be broken or discontinuous in some way the most probable explanation is that another figure is standing in the way.

For this reason, we should regard perceptions not as pictures of reality but as hypotheses about it, and as with all hypotheses the conviction with which it is held depends on the number of clues in its favour. For instance, one can destroy the Kanizsa illusion altogether by displaying the broken triangle on its own, for although such a figure is compatible with the existence of an overlaid white triangle there are no additional cues to weight the probability in this direction. But as soon as one places three black dots opposite each of the three breaks, the white triangle springs into subjective existence, although the illusion is considerably less vivid than it is with the three sectored circles.

Sometimes it is possible to create figures where there are two equally probable interpretations of what is presented to the retina. In 1832 the Scottish physicist Sir David Brewster published an account sent to him by the Swiss naturalist Necker describing how the picture of a simple rhomboid may be seen to alternate in depth. Necker found that when the figure was stripped of any accessory information which might weight the perceptual probability one way rather than another the subject was unable to make a decision. But in vacillating between these two there was never any doubt that there was a three-dimensional object in the offing, in spite of the fact that the pattern displayed on the retina was nothing more than a two-dimensional arrangement of intersecting lines. So in both cases what was seen has non-optical characteristics of tactile solidity. The mind had created a fictional reality, or rather a pair of competitive fictions which went far beyond the infor-

If one takes Descartes's image literally one would have to regard the immaterial soul as a miniature person who operated inside the pineal gland like a phantom signalman. But who tells the signalman what to do? Another signalman inside him?

mation provided. It was not seeing what had been given but perceiving a model which included what was given. In fact, according to the British psychologist Richard Gregory the sensory input provides symbols or clues which enable the brain to select among its alternative models of reality.

This is not altogether surprising when one considers the biological function of perception. The purpose of perception is to guarantee as far as possible that all actions pay off, and although certain behaviour can be successfully performed without the intervention of conscious experience, by means of reflexes which are pre-set to respond whenever an appropriate pattern of stimulation is supplied, the more elaborate strategies of living creatures depend on some process which intervenes between sensory input and motor output: some process or processes which go beyond the mere survey and recognition of incoming patterns. In contrast to the information which the senses provide, the intelligent brain reproduces not patterns but fictional objects, since it is in and amongst a world of objects that behaviour takes place. If the creature was tied to the unelaborated input of its own senses it would be unable to plan forward-looking strategies, since what was seen would be constantly changing its appearance, and there would be no guarantee that the actions which proved fruitful in the face of one lot of appearances would

What one sees depends to a large extent on what one assumes there is. Abraham Lincoln is at least one of the possibilities.

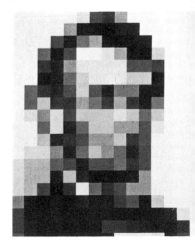

The living brain learns to see a single voluminous object from its several flat appearances. The recognisability of any one presupposes a memory of all the others.

prove equally fruitful once things had moved on. A cup seen from one angle might be regarded as something altogether different when seen from a different viewpoint, whereas the provisional theory that there is a changeless object with a repertoire of alternative aspects allows one to approach it with equal confidence from any direction. Similarly, by reading off the non-optical characteristics of slipperiness from wavy gleams one can decide to move carefully in advance without having to undergo the risks of testing the slipperiness directly. In other words, these brain fictions confer momentum and verve upon behaviour.

There are, however, obvious risks in basing behaviour on theoretical models — to plan one's future on the assumption that it will have the same structure as the past may save labour but it can also cost life. But since the results of behaviour are invariably fed back into the system which initiated it, the fictions are automatically updated by the success or failure of the actions to which they gave rise.

The simplest and most elegant way of demonstrating the effect of cerebral fictions is to use the tracking technique which became popular amongst those experimental psychologists who had worked on gun aiming during the Second World War. By using a joystick, subjects can track a moving spot on a television set with a follow spot of their own. If the target spot is made to move with a regular rhythm either up and down or from side to side, the subject soon generalises the mathematical characteristics of the moving target, and within a few cycles he will continue to reproduce the target's trajectory even when the experimentalist introduces deliberate gaps in the target display. But the system remains vigilant in spite of the success of its fiction and within a few seconds of the target changing its behaviour the tracking follows suit.

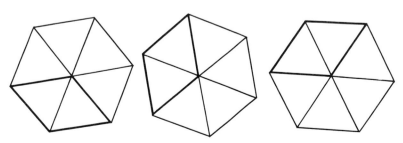

This simplified demonstration represents what is going on from moment to moment in normal perception. Behaviour often leaps quite successfully across long gaps in the sensory input, but when the person or creature comes a cropper and the fiction is found to be an unreliable representation of reality,

With the devices he has invented for managing and mastering the physical world Man has inadvertently created models of his own action. By inventing instruments which automate calculation he has also reproduced replicas of the very thought processes with which he seeks to understand that world. The invention of the micro-chip has revolutionised manufacture of computers. With this sort of technology it may well be possible to reproduce human intelligence in a machine which is not necessarily larger than the human brain. Would we recognise the humanity of such machines? Would we grant them civil rights?

the model is automatically refashioned in order to take account of the accident.

In the last two decades scientists and philosophers have noticed the striking similarity between the way in which the brain spontaneously perceives reality and the way in which the scientist deliberately explains it. Nowadays, the process of perception and the method of science are both seen to depend on making creative conjectures about the nature of reality, and upon testing and remodelling these fictions until they self-evidently coincide with the outlines of the world's facts. In science observation would dwindle into haphazard impotence unless guided by a theory which determines what will count as a relevant fact. In the same way, perception would be blind without a background of fiction to sharpen and direct its curiosity. Natural science is the native idiom of the brain made explicit. With the circular generosity of the Three Graces, the human brain has handed back to its owner the most fruitful method of elucidating its own nature.

(Correggio, 'Three Graces', Camera di S. Paolo, Parma)

Bibliography

It would be tiresome and pretentious to dangle a long bibliography on the end of a book like this, but readers who wish to pursue the subjects I have mentioned in greater detail might find some of the following texts useful and interesting.

I General works of reference on the philosophy of science and the history of medicine

Max Black, *Models and Metaphors* (Cornell University Press, 1962).

E. E. Evans-Pritchard, *Witchcraft, Oracles and Magic among the Azande* (Oxford University Press, 1937).

Michael Foster, *Lectures on the History of Physiology* (Cambridge University Press, 1901).

N. R. Hanson, *Patterns of Discovery* (Cambridge University Press, 1958).

T. S. Kuhn, *The Structure of Scientific Revolutions* (Chicago University Press, 1970).

G. E. R. Lloyd, *Early Greek Science: Thales to Aristotle* (Chatto & Windus, 1970). *Greek Science after Aristotle* (Chatto & Windus, 1972).

P. B. Medawar, *The Art of the Soluble* (Methuen, 1967).

Everett Mendelsohn, *Heat and Life* (Harvard University Press, 1964).

Karl R. Popper, *Conjectures and Refutations* (Basic Books, 1962).

Margery Purver, *The Royal Society: Concept and Creation* (Routledge & Kegan Paul, 1967).

Paolo Rossi, *Philosophy, Technology and the Arts in the Early Modern Era* (Harper & Row, 1970).

R. Shryock, *The Development of Modern Medicine* (Oxford University Press, 1936).

Charles Webster, *The Great Instauration* (Duckworth, 1976).

II Books on perception, sensation and the body image

D. M. Armstrong, *Bodily Sensations* (Routledge & Kegan Paul, 1962).

Macdonald Critchley, *The Parietal Lobes* (Hafner Publishing, 1953).

Arthur C. Danto, *The Analytic Philosophy of Action* (Cambridge University Press, 1973).

James J. Gibson, *The Senses Considered as Perceptual Systems* (Allen & Unwin, 1968).

R. L. Gregory, *Concepts and Mechanisms of Perception* (Duckworth, 1974).

Stuart Hampshire, *Thought and Action* (Chatto & Windus, 1959).

Paul Schilder, *The Image and Appearance of the Human Body* (International Universities Press, 1950).

Ludwig Wittgenstein, *Philosophical Investigations* (Blackwell, 1964).

III Books on the sociology of healing

Marc Bloch, *The Royal Touch* (Routledge & Kegan Paul, 1973).

S. N. Eisenstadt (ed.), *Max Weber on Charisma and Institution Building* (Chicago University Press, 1968).

J. G. Frazer, *The Golden Bough* (Macmillan, 1936).

Marjorie Hope Nicolson (ed.), *The Conway Letters* (Oxford University Press, 1930).

Pierre Janet, *Psychological Healing* (Allen & Unwin, 1925).

Ernst H. Kantorowicz, *The King's Two Bodies* (Princeton University Press, 1957).

Marcel Mauss, *General Theory of Magic* (Routledge & Kegan Paul, 1972).

Frank Podmore, *Mesmerism and Christian Science* (Methuen, 1909).

Max Weber, *The Sociology of Religion* (Methuen, 1965).

IV Books on the machinery of the body

There are many popular texts on this subject and it is difficult to pick one rather than another. But those accompanying the courses for the Open University are readily available in the United Kingdom, and the various biological monographs published by *Scientific American* are the best examples of their kind.

Although it is more than forty years old, Walter B. Cannon's *Wisdom of the Body* (Norton N.Y.: Oldbourne) is still a key text, and one of the finest comprehensive surveys is given by J. Z. Young in his *Introduction to the Study of Man* (Oxford University Press, 1971). This can be supplemented by his other two surveys, *The Life of Vertebrates* (Oxford University Press, 1962) and *The Life of Mammals* (Oxford University Press, 1975).

V Specific sources quoted

Walter Bagehot, *Physics and Politics* (Beacon Press, 1956).

Robert Boyle, quoted in Gweneth Whitteridge, *William Harvey and the Circulation of the Blood* (Macdonald/Elsevier, 1971).

Realdus Columbus, *De Re Anatomica* (1559), in Michael Foster, *Lectures on the History of Physiology* (Cambridge University Press, 1901).

D. Denny-Brown (ed.), *The Selected Writings of Sir Charles Sherrington* (Hamish Hamilton, 1939).

René Descartes, *The Treatise of Man* (1662), quoted in Michael Foster, *Lectures on the History of Physiology* (Cambridge University Press, 1901).

344

S. N. Eisenstadt (ed.), *Max Weber on Charisma and Institution Building* (Chicago University Press, 1968).

Valentine Greatrakes, in Marjorie Hope Nicolson (ed.), *The Conway Letters* (Oxford University Press, 1930).

William Harvey, *On the Generation of Animals* (1651), in *The Collected Works of William Harvey* (1847), translated by R. Willis (Johnson Reprints, 1965).

William Harvey, *De Motu Cordis* (1628), translated by Kenneth J. Franklin as *Movement of the Heart and Blood in Animals* (Blackwell, 1957).

Elizabeth G. Holt, *A Documentary History of Art* (Anchor, 1957).

John Hughlings Jackson, in James Taylor (ed.), *The Selected Writings of John Hughlings Jackson* (Staples Press, 1958).

T. S. Kuhn, *The Copernican Revolution* (Harvard University Press, 1957).

Levinus Lemnius, *The Touchstone of Complexions*, translated by T. Newton (1565).

Edmund Plowden, *Commentaries or Reports* (1816), in Ernst H. Kantorowicz, *The King's Two Bodies* (Princeton University Press, 1957).

Rudolf Virchow, *Cellular Pathology* (Berlin, 1858).

Index

Credits

The author and publishers would like to thank Blackwell Scientific Publications for permission to use extracts from *Movement of the Heart and Brain* (William Harvey: *De Motu Cordis*) translated by Kenneth J. Franklin; and Oxford University Press for extracts from *Conway Letters, 1642–1684* edited by Marjorie Hope Nicolson.

They also wish to thank the following for permission to reproduce illustrations: A.D.A.G.P. Paris (©) 1978), M. Duchamp, 'Nude descending a Staircase', p. 120; Aerofilms, p. 165; Alte Pinakothek, Munich, pp. 66, 71, 243 top left; A. J. Barker, p. 234; B.B.C., pp. 24, 91, 119, 132, 133, 218, 219, 232 foot; Brian Bell, pp. 58 right, 183; Bibliothèque Nationale, Paris, p. 60 right; Bildarchiv Preussischer Kulturbesitz, Berlin, pp. 146, 227, 232 top; Blocpix®, p. 221; Brian Bracegirdle, pp. 128 foot, 209 foot, 214, 217, 220 top, 224 top, 265 top and centre, 298, 305, 310 right; the Brighton and Hove Engineerium, pp. 104, 139; British Aerospace Dynamics Group, p. 328; the British Museum, pp. 70, 147; British Piano Museum, Brentford, p. 263 left; Professor George Burden, p. 289 top left, top centre, centre; Cambridge University Library, p. 59; Cambridge University, Physics Laboratory, p. 321; Camera Press, p. 31 top by Tom Blau, p. 316 by Bo Dahlin, p. 51 by Jack Esten, p. 31 foot by Ray Hamilton, p. 237 by Frank Herrman, p. 12 by Gerard Schachmes; Casa Editrice Unedi, *La Cupola del Brunelleschi*, p. 149 right; Central Office of Information, p. 213; Central Press Photos, p. 61 right; Chicago University Press, S. L. Polyak: *The Retina: The Structure of the Retina and the Visual Perception of Space* (1941), p. 334 foot; Martin Clift, pp. 90, 94, 174, 200 right; Colorific, pp. 115, 136 foot, p. 32 by Terence Le Goubin, p. 178 by John Moss, p. 114 by M. Weissman, Colorific and *Life* Magazine, p. 110 by Ralph Crane, p. 276 by Fritz Goro, p. 329 by Yale Joel, p. 112 foot by John Sadovy; Colorsport, p. 107 foot; Julius H. Comroe, California University, p. 164; Gene Cox, pp. 117 centre, 166 top right, 204, 209 top, 231 right, 268, 272, 278 foot, 334 top; Daily Telegraph Colour Library, pp. 107 top right, 136 top, 166 foot, 265 foot left, 289 foot, 314 top, 332; Dover Publications, Eadweard Muybridge: *The Human Figure in Motion* (1955), pp. 290, 309 top; Dulwich College Picture Gallery, London, p. 243 foot right; Electronic Systems Laboratory, Massachusetts Institute of Technology, pp. 260, 274; Mary Evans Picture Library, pp. 68, 102 left, 125, 144, 163 right, 255 left, 271 top; Nicholas Evans-Pritchard, p. 86; Fogg Art Museum, Harvard University (bequest of Meta and Paul J. Sachs), p. 198 top; Ford Motor Company, p. 233; Foundation Johan Maurits Van Nassau, Mauritshuis, The Hague, pp. 74, 243 top centre and centre left; the Frick Collection, New York, p. 243 foot left; Christina Gascoigne, p. 140; Fay Godwin, pp. 142, 249; Henry Grant Photos, pp. 307 right, 324, 338 left; Robert Haas, p. 289 top right; John Hadland (Photographic Instrumentation) Ltd, p. 331 left; Leon D. Harmon and Bell Labs, p. 338 right; William Heinemann Medical Books, forthcoming publication (c. 1979/80) *Scientific Foundations of Respiratory Medicine* edited by Professor J. G. Scadding and Dr Gordon Cumming, from the chapter by Dr Bryan Corrin 'Cellular Constituents of the Lung', p. 129 top; Peter Hirst-Smith, pp. 44 left, 85, 97, 124 foot, 179, 238, 239 foot, 241, 257, 264 left, 326, 339 left; Jacqueline Hyde, p. 18; Imperial War Museum, p. 327 foot; International Museum of Photography, Rochester, New York, p. 336 top right; Iveagh Bequest, Kenwood (G.L.C.), p. 243 foot centre; Bob Jackson (© 1963), *Dallas Times Herald*, p. 112 top; Dr Jacobovitz-Derks, Hôpital St Pierre, Brussels, p. 117 foot left; Kaiser Porcelain, p. 223; André Kertèsz, p. 19; Kunglia Biblioteket, Stockholm, p. 193 left; Kunsthistorisches Museum, Vienna, pp. 226, 243 centre middle; London Transport, p. 239 top (this is not an accurate colour reproduction of the London Underground map); McGraw-Hill Book Company, John C. Eccles: *The Understanding of the Brain* (1973), fig. 1–6A taken from Mountcastle, 1966, fig. 1–6B from H. K. Hartline, *J. Cell. Comp. Physiol.*, 5:229 (1934), p. 303; the Mansell Collection, pp. 54, 61 left, 73, 76, 77, 82 top, 102 right, 103, 111, 148 foot, 254, 262 left, 263 right; S.&O. Mathews, p. 259 foot; Medical Research Council Applied Psychology Unit, Cambridge, p. 327 top; Methuen, Hugh B. Cott, *Adaptive Coloration in Animals*, fig. 18.4, p. 42 top; Metropolitan